T0229482

Bringing a Medical Device to the Market

Bringing a Medical Device to the Market
to the Market
A Scientist's Perspective

Gennadi Saiko

JENNY STANFORD
PUBLISHING

Published by

Jenny Stanford Publishing Pte. Ltd.
101 Thomson Road
#06-01, United Square
Singapore 307591

Email: editorial@jennystanford.com
Web: www.jennystanford.com

British Library Cataloguing-in-Publication Data
A catalogue record for this book is available from the British Library.

Bringing a Medical Device to the Market: A Scientist's Perspective

ISBN 978-981-4968-25-6 (Hardcover)
ISBN 978-1-003-31221-5 (eBook)

Contents

Preface

People often refer a high-level view as one that is at 30,000 feet. When one is coasting at a high altitude in a plane, it is fun to see the Earth down there. One can see large expanses of land, rivers, valleys, and high mountains. However, conditions outside the plane at this altitude are harsh—temperatures minus 50–60 °C and very low pressure. People typically do not jump from such altitudes as it can be pretty dangerous.

The average altitude for skydiving in the United States, and common to most Cessna aircraft drop zones, is 10,000–13,000 feet for routine jumps. From 10,000 feet, one can enjoy a 30-second free-fall and from 13,000 feet, one can enjoy a 60-second free-fall. Military paratroopers do not enjoy a free-fall at all. Unless it is a HALO (high-altitude, low-opening) jump, paratroopers typically jump from 1250–800 feet. In combat, they will jump from 500 or even 400 feet.

I would compare an entrepreneur's journey to a parachute jump, often on an unfamiliar territory. The first jump is always a challenge. Confidence comes with experience. It is also all about preparation. One will spend a lot of time preparing for the jump.

This book presents a 10,000 feet overview of MedTech entrepreneurship. It provides an overview of areas that I wish I knew before my first jump into this field. From 30,000 feet, one can't see the details. The book presents details to grasp the topic's complexity and gives ideas where to start and what needs to be taken care of.

Entrepreneurship as a minefield is one of the metaphors that is used often. I disagree with such a comparison. Entrepreneurial errors are costly, but their outcomes oftentimes are not that binary and not that dramatic. In many cases, one can undo the harm caused by a suboptimal decision. However, it definitely impedes one's speed and overall progress. Thus, I would compare MedTech entrepreneurship with a race on a rugged terrain. A suboptimal decision may slow one down and cost prize money at the finish.

I wrote this book for first-time MedTech entrepreneurs. MedTech's definition typically encompasses at least two areas: medical devices and *in vitro* diagnostics (IVD). Sometimes, medical devices are split into several categories, such as surgical, and many times digital health (or HealthTech) is also included in MedTech. The book is mainly for medical devices and diagnostics entrepreneurs. However, most information can be helpful for digital health entrepreneurs as well.

The book is based mainly on personal experience, the wisdom of several experts sifted at multiple conferences, seminars, workshops, and webinars, and information mined from regulatory websites. A recurrent theme in the book is the application of a scientific method. While scientists do routinely apply it in scientific research, one of the book's central premise is the necessity to apply it in many other aspects of entrepreneurship. Formulating hypotheses and verifying them in experiments should be used routinely in many product management and business aspects.

While the purpose of this book is to elucidate MedTech-specific areas, I make some incursions into universal startup areas to highlight some critical points. However, I have not supplanted the topics that have been covered in multiple excellent resources on startups.

With this book, I aim for several goals. In the grand scheme of things, I try to eliminate the knowledge gap, which exists in MedTech. I hope this book will further drive healthcare innovations and help turn the quantity of innovations into quality changes, which is long overdue.

On an individual reader level, my goal is less ambitious. When I attend any webinar or seminar, I set my expectations that if I hear at least one new helpful idea or concept, my time will be well spent. I set similar expectations for this book: if a reader learns at least several new tricks, my mission is accomplished. Each of these tricks may help save a lot of time and many thousands of dollars of new entrepreneurs.

I am extremely grateful to my wife, Vera, for her continuous support during the writing of this book. I am thankful to my son, Daniil Saiko, CFA, for reviewing several chapters relevant to his expertise.

Gennadi Saiko
Mississauga, ON, Canada
March 2022

Chapter 1

Introduction

Do you know how venture capitalists (VCs) make money? It is a numbers game. Out of 10 companies in their portfolio, just one will be a moonshot. One to two companies will return 5–10× (five-ten times more than the initial investment), another one to two companies will return 1–2×, and others just will disappear…

The moonshots' stories are typically well-publicized; you heard them a lot and knew them pretty well. My company got a "B" on this scorecard. As you see, it is a much more probable but still positive outcome.

A little bit about me: I have a pretty diverse experience starting from theoretical physics (condensed matter) to applied physics (biospectroscopy) to engineering (bioimaging) with a detour in finance.

In 2013, I had called my quite successful financial industry career off. I went as a postdoc to Ryerson University, where I started working on multispectral imaging modalities with a primary emphasis on wound care. In 2015, to commercialize the idea, I founded Oxilight Inc, which received its first funding (grants) in 2016. In 2017, we presented our prototype (a handheld device) at Symposium on Advanced Wound Care (SAWC), a major industry event in wound care. Based on our participation at SAWC, we decided to miniaturize the device. Four months later, we pivoted a small attachment to

Bringing a Medical Device to the Market: A Scientist's Perspective
Gennadi Saiko
Copyright © 2022 Jenny Stanford Publishing Pte. Ltd.
ISBN 978-981-4968-25-6 (Hardcover), 978-1-003-31221-5 (eBook)
www.jennystanford.com

a smartphone. In 2019, I sold my technology to Swift Medical, a prominent digital wound care space player.

Entrepreneurship and startups are hot topics these days. There are a lot of good books about them. Why do we need another book about startups? This book is different (you probably heard it before☺). It is about medical technology (MedTech) startups, which are very different from others. Regulations are the primary reason why they are different. While there are other regulated industries (e.g., FinTech), healthcare regulations have a much more profound impact on the startup's trajectory. This entire book is about why MedTech startups are different, how they are different, and what to do about it. In a nutshell, it is a manual for a future MedTech entrepreneur.

In the MedTech world, if you invented something, basically, there are two primary ways to commercialize your idea. You either can go to your Technology Transfer Office (aka licensing) or commercialize your idea by yourself (aka startup). Both of these methods have their own pros and cons. We will briefly touch on the licensing path; however, this book's primary topic is commercialization via startup.

The chances of failure for startups in general and MedTech, in particular, are very high. However, there are approaches on how to increase the chances for survival and success. This book attempts to gather and present all information relevant to a MedTech entrepreneur in one place. I tried to avoid discussing general startup topics, so you can consider this book as a MedTech add-on.

1.1 Startups

The startup is a pretty broad term, which is used in different contexts. It can be helpful to establish some common language.

We will use a taxonomy of startups and early-stage companies based on their ability to scale perceived by investors. In this case, the lifecycle of the startup can be split into three major phases:

- A startup phase (pre-Series A company) is defined by an effort to de-risk. The question is whether or not the company could scale. This phase includes multiple stages (initial idea, pre-seed, seed stage, etc.)

- A scale company (Series A company): The Series A round takes place when professional investors are convinced that a company could scale
- A growth company (post-Series A company)

Thus, Series A is a significant milestone from the funding and survivorship perspective.

Pre- and post-Series A phases can be further split into subphases. Generally, a "startup" refers to a very early stage of the company. However, what is the "early stage"? "Early stage" typically refers to the state where most aspects of the company remain incomplete, although there is evidence of progress in its development. The early stage can have several subphases, separated by one or more funding rounds (seed round), typically from company founders, friends, or family members.

Occasionally, "startup" refers to a business, which has been on the market for less than three years. However, this is not accurate, particularly for MedDev companies. Its progress, not its age, defines the company's stage; however, the age is still relevant. If the company has not made much progress within the first three years, it definitely loses its curb appeal from a funding perspective.

Post-Series A phase can also be split into early growth (Series A and B), late growth (Series C+), etc.

Thus, we will refer to a startup as a company in the early stages of development, which is in the business to solve real-life problems through a product or innovative service.

Now, let's consider the major types of startups. It will help us to define the range of possible outcomes. According to Steve Blank, there are six different types of startups:[1]

- **Lifestyle Startups** are founded by entrepreneurs who are working for themselves on what else they like. Examples of these are freelancers or Web designers who have a passion for their work.
- **Small Business Startups**, where the owner follows less ambitious goals to provide only a comfortable life for his family. Examples of these are hairdressing salons, grocery stores, bakeries, among others.

---

[1]Blank S (2013): The 6 Types of Startups, *WSJ*, Retrieved from https://www.wsj.com/articles/BL-232B-1094.

- **Scalable Startups** are founded by entrepreneurs who believe from the beginning that they can change the world with their business idea and therefore worry about finding a model scalable and repeatable business to draw investors' attention to boost your business. Examples of these are Google, Uber, and Facebook.
- **Buyable Startups**. These businesses are born with the goal of being sold to large companies after achieving positive results that catch their attention. This type of startups is particularly common in Web solutions development companies and mobile. An example of this was the purchase of Instagram by Facebook.
- **Large Company Startups**. These businesses have the main objective of innovation and have a limited duration of life. These businesses develop products or services that revolutionaries become quickly recognized by the market. However, due to market changes, user preferences, competitive pressures, these businesses tend to create new innovative products for new users of different markets.
- **Social Startups**. Finally, these entrepreneurs want to make a difference in society and make a better world. Their focus is not to gain profit but rather to contribute positively to the community. One example is a charity or charitable institution.

So, which options are relevant to a MedTech entrepreneur? First, we can quickly eliminate the large company startup option. We are now left with five options.

We know that there are successful non-profits in the healthcare services space (for example, Saint Elizabeth Health). We also know that multiple companies try to bring cheaper medical solutions to poorer countries. However, social startups can be quite a complicated case in MedDev because social startups are run as non-profits. Thus, sources of funding will be quite different from for-profits. Instead of private capital, it must rely on other sources as government grants and fundraising. We will cover this more extensively later in the book, but the time to market is quite long due to extensive regulations. Thus, unless it is an unregulated digital health project, you will deal with a long-term and capital-intensive project with very limited financing options. This mismatch between a long-time horizon and limited funding sources will create undue strain on the already strenuous process. So, unless you secured multi-year funding from a government or philanthropic organization beforehand or found a way to do it unregulated, social startups will be an extremely challenging option in MedTech.

While the cost of running a business significantly dropped in recent years; still a MedTech company needs to be (for reasons explained later) well-capitalized. Lifestyle and small business companies have limited sources of funding. Thus, they can be a viable option only if you have almost unlimited access to capital, which certainly is not always the case. Therefore, it leaves us only with two options: scalable startups and buyable startups.

Consequently, the positive outcome (an exit) for a MedTech entrepreneur is either a merger and acquisition (M&A) (buyable startup) or initial public offering (IPO) (scalable startup). We can rank these opportunities even further if we consider the corporate deal statistics in the MedDev space. Prior to 2010, statistics were skewed toward M&As: more than 90% of corporate transactions in the MedDev space were M&As. However, most current data shows that this ratio is close to 2:1 now (see Table 1.1).

Table 1.1 Private M&A and IPO in medical devices space (global data for 2014–2020)[2]

Year	Private M&A	IPO
2014	18	10
2015	19	11
2016	12	3
2017	14	3
2018	20	8
2019	17	8
2020	16	11

Thus, we can draft our high-level plan as the following (see Fig. 1.1). We have an idea at the beginning and several potential endgames (exits): through licensing, IPO, or M&A

[2]Silicon Valley Bank (2021), Healthcare Investments & Exits Annual 2021 Report, Retrieved from https://www.svb.com/globalassets/library/managedassets/pdfs/healthcare-report-2021-annual.pdf

What we need to do next is to connect the dots between these two states. Following a systematic approach does not guarantee success; however, it reduces the chances of failure.

Figure 1.1   A high-level plan for MedTech entrepreneur.

So, where can we start? The most obvious candidate will be to file a patent (if it is not filed yet). However, there are certain considerations about the patent's timing and scope, which we will address later. Instead, my suggestion will be to start with customer discovery. It is of paramount importance.

1.2   Customer Discovery

You will be surprised how many great technical solutions are stored on the shelves and never or barely used. It can be devastating to spend several years of your life and a significant amount of money to learn that there is no need for your great idea or the solution is so cumbersome that the user cannot use it.

So, before you start investing a significant amount of time and money, it is better to understand your user/customer and whether they need it. The reasons are pretty simple. First, you are typically blindsided by your idea, so you need an independent opinion. Second, unless you are a healthcare professional, you are not an expert in the healthcare field and need an expert opinion from a user perspective.

Customer discovery is vital to any startup; however, it is paramount for a MedDev startup. The reason is straightforward. With software development, you can do multiple releases per day and pivot a new solution every week or month. In hardware development (where most MedDev startups are), you are typically limited to 1–2 releases per year. Thus, pivoting is much more complicated. However, the MedDev world is the next level of complexity because it a heavily regulated industry. Thus, if you want to pivot something, it may take years. Therefore, you virtually don't have room for error. It would be best if you did it right the first time. The goal of customer discovery is to reduce the risk of product/market misfit. The cost of amending the course rises with time spent. Setting the right path early saves a lot of time and money. And customer discovery is the right tool for that.

While customer discovery sounds scary, fortunately, there are simple ways to do it. You need to run customer interviews.

The idea behind customer discovery is to find a problem which you are trying to solve. During customer interviews, you need to understand your potential customer's problems and whether the pain you want to solve is on the customer priority list.

So, basically, your potential customer needs to answer the following questions:

- What problem do they have?
- How do they solve it today?
- What is not ideal about their current solution?

At this stage, it should be an open question discovery. Do not pivot your solution. It is all about customer problems, not your solution.

How many customer interviews do you need? As a rule of thumb, 30 interviews will start giving you a statistically meaningful answer. However, in some niche markets, it seems like an almost insurmountable task. So, probably you do not need to go that far. Even after several interviews, you will start seeing common themes. So, probably just five good-quality interviews will be sufficient for customer discovery purposes.

I want to repeat again that customer discovery is of paramount importance for any MedTech startup. You cannot skip this step. However, customer discovery in healthcare is no different from any other field (other than access to healthcare professionals, which can be challenging, especially during pandemic times). Thus, you can use a vast body of knowledge developed by entrepreneurs, product managers, UX research professionals to do it properly.

Also, note that you cannot delegate customer discovery to anybody else. Founders or members of the founding team have to do it by themselves.

Customer interviews will help you understand the breadth of the problem you try to solve and can be an essential input into a market size assessment, which you will do next.

1.3 Market Size

At this stage, you have to understand the problem and who has this problem. Based on this understanding, you need to run a preliminary market analysis. There is no need to buy expensive market research. All you need to do is to estimate the market very roughly. For example, how many surgical procedures of this type have been performed annually? Or how many hospital beds are and how many beds per your device are you aiming for?

From these numbers, you will be able to estimate the total market size. It is a very important number, which can determine your entire future trajectory. Is the problem worth your time and energy? Will this problem be of interest to investors? We will briefly discuss this topic in Chapter 9 (Other Considerations).

1.4 Patent Filing

The first US patent was granted on July 31, 1790. Earlier that year, the US Congress passed the first patent act titled "An act to promote the progress of useful arts." As stated in the title, the original idea behind the patent system was to promote sharing information. It encouraged sharing information in exchange for some protection to the inventor.

Since then, the patent system almost completely lost its original aim of sharing information; however, it is still a pillar of intellectual property (IP) protection.

This book is not about IP laws; however, it can be helpful to provide some basic information about IP.

There are four forms of IP (patents, trademarks, copyrights, and trade secrets), which can be legally protected in most countries.

A patent is an IP form, which most people are familiar with. It gives protection for 20 years since an original filing (priority date). The idea behind a patent is to create a negative space where your competitors cannot operate. Your goal is to expand this negative space as much as you could.

Patents do not provide IP protection cross-border. They need to be filed in each country where you are seeking IP protection. And each filing is quite costly. So, historically, you had to spend a lot of money if you wanted to protect your invention. However, filing patents does not guarantee commercial success. Most of the patents filed have never been used by anybody and have not generated a single penny. Thus, you had to spend quite a significant upfront cost by filing a patent application without any guaranteed return.

Fortunately, in 1970, with the Patent Cooperation Treaty (PCT) signing, the situation changed. PCT allows seeking IP protection simultaneously in most countries (including all industrialized) by filing a so-called PCT application. It does not eliminate the necessity of national phase filings. However, it defers these significant costs for 30 months (if combined with a provisional patent). So, for a fraction of the cost of national phase filing, the PCT route gives you 2.5 years to decide whether it is worth incurring these costs and which markets are important for you.

If you follow the PCT route, your patent application will be published and become publicly available after 18 months since the original filing. Moreover, during the PCT filing, you can elaborate on existing ideas or add new ones. If you come up with new ideas, then they will be given a newer priority date.

These considerations can help you to time the patent filing. You need to consider the scope of claims (which may increase in time with new data and ideas) and your readiness to execute your plan. You have only 18 months before your idea becomes publicly available and only 30 months before you need to spend significant money during the national phase.

To understand your IP position, you may need to perform a freedom-to-operate (FTO) search. FTO's goal is to assess your ability to produce and market products without legal liabilities to third parties.

FTO can be performed internally (typically called a preliminary patent search) or by a patent lawyer. In the latter case, you can additionally request a freedom-to-operate opinion.

There are multiple pros and cons of obtaining the FTO opinion. It will definitely strengthen your case when you talk with investors. However, the biggest shortcoming of any significant effort in the FTO is its potential incompleteness. As we mentioned earlier, under the PCT route, the patent application is not publicly available for the first 18 months. Thus, newly filed applications will not be captured in an FTO or preliminary search.

However, despite this shortcoming, there are still multiple good reasons to internally perform the preliminary patent search. First, you may find some competitors. Second, if you find them, you will see whether you can craft your claims to bypass them. Finally, the patent language is quite particular, so it is good to start getting familiar with it.

The preliminary search can be as simple as a search using Google Patents or free online resources of the United States Patent and Trademark Office (USPTO) or the European Patent Office (EPO).

So, you have an idea (and hopefully some data, which supports it), user feedback and market analysis, and data from the preliminary patent search. Now, you are in an excellent position to start crafting your patent application. Again, you can do it internally or engage a patent agent. You can also prepare the draft by yourself and get it revised/amended by a patent agent.

In many cases, the original filing (e.g., provisional patent application in the United States) can be as simple as a journal paper with an additional claims section.

Crafting good claims is a particular skill, so it is a good idea to get them verified, especially if you are a first-time inventor.

One important thing to remember about your idea's patentability is public disclosure (e.g., a publication, conference, or seminar). The US and Canadian laws give a year grace period for public disclosure, which means that you can file a provisional patent application within one year after the first public disclosure. However, most of other countries do not have such a grace period.

1.5   Paths to Commercialize

Now, it is also an excellent time to decide what to do with your invention. The decision point here is whether to license it or to commercialize it by yourself. At this point, you are well equipped to make this determination. You have a customer's input, an initial understanding of market potential, and a rough idea of whether you have IP. However, there are more factors, which will affect the success of your project. And they need to be taken into consideration. One of the most critical factors here will be the medical device class from a regulatory perspective. We will address this topic in more detail later; however, in a nutshell, the medical device class corresponds to the risks associated with a particular device. For example, there are three regulatory classes in the United States: Class I (with and without exemptions), Class II (with and without exemptions), and Class III. The class determines, among other things, the type of premarketing submission/application required for FDA clearance to market. If certain Class I and Class II devices can be exempt from notifications, a Premarket Approval (PMA) application will be required for Class III devices.

This class determination gives an indication of the amount of effort, which needs to be done to get the device on the market. It also gives you a clue about how much money do you need. Market professionals' rule of thumb for the just product development costs are $1Mln for the Class II device and $10Mln for the Class III device. As you can see, the costs of Class II and Class III devices are incomparable, and the amount of regulatory scrutiny will be completely different as well. Clinical trials (if any) and regulatory costs are also quite significant, particularly for Class III devices. So, most likely, you will need 10× as much money (see Chapter 5 for details).

So, now you can finally decide on the pathway. If you see the market potential, the market is big enough, and know how to get the required money, then the startup pathway is feasible. If you are unsure about market potential and/or where to get the necessary capital, licensing can be a safer option.

Last but not least is a passion for your invention and the problem you try to solve. Without deep routed passion and motivation, it will be hard emotionally and materially to succeed, as there will be many ups and downs. Commercialization through startup is a long and enduring process. If you are really fervent about solving this problem, then the startup pathway can be for you. If you want to do it in parallel with many other things, then think twice before jumping into the startup world.

1.6   Technology Transfer Office

If you made your invention in a university or hospital, your institution might have a stake in your invention irrespective of how you commercialize it. Universities and hospitals typically have a Technology Transfer Office or something similar (e.g., the Office of Technology Licensing at Stanford University), which takes care of the organization's IP portfolio. The exact share, which will belong to the institution, depends on the institution's policies. It varies significantly between hospitals and universities. For example, some universities (Stanford University in the US, Ryerson, Waterloo, and Simon Frazer Universities in Canada) have quite liberal policies on owning IP developed in the university. Liberal policies promote commercialization. Stanford University's IP policies are exemplary and one of the keystones of Silicon Valley's success. At the time of writing, Stanford's policy on royalty sharing[3] includes "net cash royalties are divided 1/3 to the Inventor, 1/3 to the Inventor's department and 1/3 to the Inventor's school." Other universities may take a more significant stake in inventions.

   The size of the stake also may depend on IP services provided by the institution. It can be lower if you take care of patent drafting and filing. However, it can be higher if the university will perform these functions. The university's stake can be even higher if it is responsible for the marketing of the invention. The significant advantage of licensing the patent through the Technology Transfer Office is its ability to support the whole licensing process. They have incentives, experience, and all means to negotiate the external deal fairly. Licensing through the university (particularly through a famous university) also gives a lot of credibility to your invention.

---

[3]Stanford Policies, Stanford University, Retrieved from https://otl.stanford.edu/intellectual-property/stanford-policies

So, it is vital to understand the university's IP policies and engage the Technology Transfer Office early in the process. They will be able to help and guide you through many intricacies of the IP process. Moreover, it can be your obligation to disclose the potentially patentable invention, which was developed (conceived or reduced to practice) by faculty or staff of the university in the course of their university responsibilities or with more than incidental use of university resources, to the university on a timely basis.

There are multiple ways to set up a legal structure around IP depending on the school policies. For example, in some cases, the university owns the invention's title, and it needs to be licensed to the inventor, which wants to pursue commercialization through the startup. Other universities may allow setting up a corporation, and the university will be a stakeholder in such a corporation.

Note that while it is wildly expected that the university will take active steps to find a licensee for your invention and you can sit and relax, in reality, it is not always the case. There are much more chances that a potential licensee finds you rather than the invention in the university archives. So, to get it licensed, you have to promote your invention actively by yourself. Fortunately, there are multiple avenues to do it, including conferences, trade shows, and publications. Once you found the licensee, the university will be happy to negotiate and close the deal, which is very important by itself.

Another aspect to consider if you select the licensing pathway is the type of license: exclusive or non-exclusive. While it is tempting to license your invention to multiple companies, it may have a trap. If the company wants to develop and bring to market a breakthrough technology, the last thing they want is to bring it simultaneously with two or three of their competitors. Thus, it may slow down negotiations and reduce their commitment, ultimately resulting in fewer royalties.

Hopefully, you decided to commercialize your idea, considered licensing and startup, and selected the startup pathway.

The goal of this book is to help you to get to the first professional investment round. After that, you probably will be in experienced hands, which will help you achieve further targets. However, to achieve this goal, you have to think and work strategically.

The rest of the book will cover all necessary significant steps you need to encounter and some factors, which are essential for success.

The book consists of two parts. Part 1 (Road to Market) is about necessary steps, and Part 2 (Important Ingredients) is about critical factors for success.

In Part 1, we will discuss the overall road map of bringing a medical device to the market and briefly discussing all necessary steps.

However, we will start working backward on this process and focus on the second last step of this process: regulatory approval or clearance. While the final step, Go to Market, is also very important; however, regulatory clearance is paramount. Once you have the product cleared, you have much better chances to get funded. Not to mention that you may start marketing and selling your device. Why we focus on regulatory clearance first? Medical devices is a highly regulated industry. It distinguishes it from virtually every other market. While consumer goods, for example, also require some certifications, their complexity is not even close to MedDev regulations. Thus, the necessity of regulatory clearance defines all preliminary steps. To succeed on this path, we need to set our mindset and our expectations clearly. Therefore, we start our discussion with Chapter 2 (Regulatory Environment).

Once we understand the regulatory environment, we must understand how to comply with all these regulatory requirements. Chapter 3 (Prerequisites) will discuss quality management, risk management, and EMC/safety certification. Here, EMC stands for electromagnetic compatibility. While this step may be required for other hardware-based products, the level of their complexity in the MedDev world is also higher. Thus, we address some basic concepts of this process.

The other necessary step to get regulatory clearance is to generate data. While EMC/Safety testing focuses on safety, you also need to perform performance testing and show that your device delivers the functionality you claim. For this purpose, your device needs to be tested in the lab and/or on patients. The level of testing varies significantly with the imposed risks and the novelty of your device. While innovations in well-established technologies may require a moderate and well-defined set of testing, ground-breaking devices will be subjected to significant scrutiny so that extensive testing will be required.

Finally, with understanding what to expect, we are ready to look at product development. Given all this regulatory complexity, product development in the MedTech world has some distinct features, and they will be discussed in Chapter 4 (Product Development Process).

We will finish Part 1 by discussing the timelines and capital required to make your venture happened. It will be discussed in Chapter 5 (Timelines and Capital).

Part 2 (Important Ingredients) of the book is devoted to the most critical factors, which affect your success.

We will start with the funding. While there are some discussions about the most crucial factor for the startup's success, the answer is evident in the MedDev world—funding. MedDev product development is a long and enduring process. You need to be well-capitalized to conquer it. We start by discussing the various sources of funding. We will then talk about which sources are available and feasible at different stages of product development. This will be the focus of Chapter 6 (Funding).

The other factors which we will discuss are your competitive advantages. We call Chapter 7 "IP and Other Moats." We will start with IP and discuss various types of IP and some specifics of the patent application process. However, there are other ways to protect your business. We will briefly discuss some other options later in that chapter.

Then, we will talk about some aspects of your business model. Chapter 8 (Business Model) will try to help you answer who will pay for your device. While a patient can incur out-of-pocket expenses, the business model, which relies on out-of-pocket expenses by healthcare professionals, can be unsustainable. Thus, we need to find another source that will pay for your device. In many cases, it can be an insurer. Some countries (Canada, UK) have a single-payer system. The other countries (e.g., the United States) have multi-payer systems. The process of payment by a health payer is called a reimbursement. We will talk about some basics of reimbursement, with particular attention being paid to the US market.

Chapter 9 (Other Considerations) will discuss other important factors that impact the outcome. The first important factor to discuss is team. While some great apps can be designed, released, and maintained by a single person, any medical device requires many people's work. Some of the work can be outsourced and performed by an outside company or a freelancer. However, some knowledge and information should stay in-house, and you need a team for that.

Then we will discuss the scale and the novelty of the idea. It also has significant implications on the overall dynamics and outcome.

This chapter will also discuss several other topics (e.g., age of entrepreneur, therapy vs. diagnostics), which are less critical; however, they are still significant.

Chapter 10 (Silver Lining) will discuss why being a scientist increases your odds of succeeding in a MedTech venture.

The well-established world has been changed recently with COVID-19 pandemic. The MedDev world has been changed, as well. We will briefly discuss its implications and future potential impacts in Instead of Afterword.

Notes to the Reader

1. Chapters 2 (Regulatory Environment) and 3 (Prerequisites) are regulatory-heavy and may be hard to read. They can be skipped or skimmed during initial reading and used as a quick reference thereafter. Also, notice that the regulatory environment is subject to constant changes. I tried to keep the content of these chapters as accurate as possible as of the time of writing (Fall 2020–Winter 2021). However, things may change rapidly and dramatically. It is advisable to research an appropriate regulatory topic before any infliction point. The regulatory information is typically available on regulators' websites. Also, all recent updates are almost immediately getting commented on by the regulatory consultant community. Thus, Google is your best friend.

2. While any MedTech project roots in science and engineering, at its core, it is still a product with a significant business component. I present here multiple approaches, which were validated in many MedTech projects. However, it is art rather than science. Experience and confidence come with practice. As such, this book is rather an opinion than a scientifically proven matter.

PART I
ROAD TO MARKET

Proper product development is of particular importance for medical devices. Following a certain methodology and roadmap is of great help during this long process and increases chances for success.

In Part 1, we will talk about the product development roadmap and its most important stages.

The MedDev product development roadmap is defined by a particular event, which is a focal point of the whole process. Regulatory clearance is such a culmination point and the major outcome of the project.

The process can be split into several phases: R&D; product development; pre-clinical, clinical, and regulatory phases.

The first phase is R&D. During this phase, a rough prototype of the device should be developed. The most important part here is to get user feedback and incorporate it into the design. Thus, more realistically, multiple generations of the prototype need to be developed and tested during the R&D phase.

When you are ready to finalize product requirements, you can move to the second phase, a product development phase. During this stage, a market-ready (or close to market-ready) product will be developed. While this phase can be seen as a continuation of R&D, it has several distinct features, making them different.

When you have a market-ready product (or product close to market readiness), you need to evaluate its performance. It can be done in the lab (pre-clinical phase) and in some cases on human subjects (clinical stage). The goal of these phases is to show the safety and efficacy of your device.

Finally, when you have your market-ready device, and you collected substantial evidence about its safety and efficacy, then you can move to the last phase, the regulatory clearance phase.

When you got your device cleared, then you can start market and sell it.

Each phase consists of multiple steps. Some of them are strictly sequential. Others are interconnected. While you can do certain steps in parallel, however, in general, the whole process is very linear and sequential. This sequentiality brings certain implications: (a) the entire project is quite long (we will get a time assessment for each phase later), and (b) it is hard to expect that you will be able to crash this schedule and deliver the project much faster.

Medical device product development is different from other hardware projects. The key difference here is the necessity of regulatory clearance, which dictates the gathering of significant evidence on safety and efficacy. Thus, the medical device development will take much longer than the identical device tailored to non-medical applications.

Medical device product development is also different from drug development. In many cases, the key difference here is that the body of evidence required to demonstrate safety and efficacy is much smaller for a medical device than for a drug. Any new drug requires three rounds of clinical trials (I, II, and III), each with increased scope and complexity. Thus, drug development typically is a much longer and much more costly project.

We will start our discussion with the regulatory phase. This phase is a defining one in the medical device journey and dictates many peculiarities of the product development roadmap.

Chapter 2

Regulatory Environment

Medical devices are diverse. They range from a simple scalpel to smartphone apps to implanted pacemakers. They include *in vitro* diagnostics (IVD) products, such as reagents, test kits, and blood glucose meters. They also include diagnostic ultrasound products, X-ray machines, and surgical lasers.

As we mentioned before, a medical device (MedDev) product is different from other hardware projects because it is necessary to comply with various regulations.

Most counties have specialized government agencies, which are responsible for protecting public health. Examples of such agencies are the Food and Drug Administration (FDA) in the United States, Health Canada in Canada, the European Commission's Directorate-General for Health and Food Safety in the European Union (EU), or the Department of Health in Australia. The mandate of these regulatory bodies, among other things, is to ensure the safety, efficacy, and security of drugs, biological products, and medical devices.

To ensure the safety, efficacy, and security of drugs, biological products, and medical devices, the government agencies regulate their marketing and sales. Medical devices are part of these regulations.

Bringing a Medical Device to the Market: A Scientist's Perspective
Gennadi Saiko
Copyright © 2022 Jenny Stanford Publishing Pte. Ltd.
ISBN 978-981-4968-25-6 (Hardcover), 978-1-003-31221-5 (eBook)
www.jennystanford.com

What precisely a "medical device" is? Let's start with the definition. A "medical device" is defined in Section 201(h) of the US's Food, Drug, and Cosmetic Act[1] as

"An instrument, apparatus, implement, machine, contrivance, implant, *in vitro* reagent, or other similar or related article, including a component part, or accessory which is:

- recognized in the official National Formulary, or the United States Pharmacopoeia, or any supplement to them,
- intended for use in the **diagnosis** of disease or other conditions, or in the **cure, mitigation, treatment,** or **prevention** of disease, in man or other animals, or
- intended to affect the structure or any function of the body of man or other animals, and
- which does not achieve its primary intended purposes through chemical action within or on the body of man or other animals and
- and which is not dependent upon being metabolized for the achievement of its primary intended purposes..."

There are several key elements of this definition. The most critical aspect here is the intended use. It should "**diagnose disease, or cure, mitigate, treat,** or **prevent** the disease." The same device (e.g., a knife or diapers) can be a medical device or not, depending on its intended use. The intended use also includes another critical element: indication of use, which can be described as "the disease or condition the device will diagnose, treat, prevent, cure or mitigate, including a description of the patient population for which the device is intended." Diapers are the classic examples here. Diapers for children are not a medical device. However, the diapers for adults to mitigate incontinence will be considered a medical device.

To allocate scarce resources more efficiently, most counties have adopted a risk-based approach to medical device classification. In this model, the amount of regulatory scrutiny depends on device risk classification. More evidence about safety and efficacy must be provided for higher-risk devices such as pacemakers or X-ray

---

[1]FDA (2017) Classification of Products as Drugs and Devices and Additional Product Classification Issues. Retrieved from https://www.fda.gov/regulatory-information/ search-fda-guidance-documents/classification-products-drugs-and-devices-and-additional-product-classification-issues

machines. Less evidence is required to demonstrate safety and efficacy for lower classification devices such as ECG machines or bandages.

While in the last years, there have been attempts to synchronize and harmonize approaches between regulators, but there are still significant differences. So, it is easier to explore it on an individual market level. We will briefly review approaches and regulatory environments in several major MedDev markets, including the US, EU, Canada. Australia, and Japan. We will begin by reviewing the US medical device market, the biggest MedDev market in the world. To show the breadth of regulations, we will discuss the US market in more detail. Other markets will be reviewed with fewer amounts of details.

2.1   US Regulatory Environment

The US has one of the world's most comprehensive and effective public health and consumer protection networks. The US MedDev market is regulated by the Food and Drug Administration (FDA). The agency enforces compliance with the Food and Drugs Act, which was originally signed into law in 1906.

FDA has a broad mandate. It regulates human and veterinary drugs, biological products, medical devices, tobacco, food safety, cosmetics, and products that emit radiation (e.g., microwaves).

Most medical devices are regulated by the Center for Devices and Radiological Health (CDRH), an FDA division. The CDRH approves and clears medical devices for marketing in the US, maintains the approved devices' database (device registration and listing database), and performs post-market surveillance. However, not all medical devices are regulated by CDRH. Devices for blood collection, blood products, cellular therapies, and specific medical devices' combinations with these biological products are regulated by the Center for Biologics, Evaluation, and Research (CBER). However, even though CBER reviews these products, medical device laws and regulations still apply.

The CDRH (CBER) sets expectations for the market participants by issuing guidance documents.

FDA uses a risk-based approach for the classification of medical devices. In 1976, the Federal Food, Drug, and Cosmetic Act established three regulatory classes for medical devices based on the degree of control necessary to ensure that various devices are safe and effective.

"Class I – These devices present minimal potential for harm to the user, subject to general controls. Examples include enema kits and elastic bandages. 47% of medical devices fall under this category, and 95% are exempt from the regulatory process.

Class II – Includes devices for which general controls are insufficient to assure safety and effectiveness. Performance standards are required. Examples of Class II devices include powered wheelchairs and some pregnancy test kits. 43% of medical devices fall under this category.

Class III – Devices with the greatest risk and where insufficient information exists to assure safety and effectiveness. These devices usually sustain or support life, are implanted, or present potential unreasonable risk of illness or injury. Examples of Class III devices include implantable pacemakers and breast implants. Approximately 10% of medical devices fall under this category." [2]

Based on the risk profile of the device, there are specific pathways to enter the US market. The three primary pathways are exempt products, Premarket Notification [510(k)], and Premarket Approval (PMA).

2.1.1   Exempt Devices

The standard way to enter the US market and start marketing the device is to file a premarket submission with the FDA. "Exempt device" means that the device is exempt from a premarket submission process. Most Class I and some Class II devices fall into this category. Examples of exempt devices are manual stethoscopes, mercury thermometers, and bedpans. A device may become exempt if the FDA decides that a 510(k) is not required to provide reasonable assurance

---

[2]FDA, Learn if a Medical Device Has Been Cleared by FDA for Marketing. Retrieved from https://www.fda.gov/medical-devices/consumers-medical-devices/learn-if-medical-device-has-been-cleared-fda-marketing

of safety and effectiveness for the device. Approximately 74% of the Class I devices are exempt from the Premarket Notification process. Devices that may be exempt from 510(k) requirements include:

- So-called "pre-amendments devices"; and
- Class I and Class II devices, which the FDA specifically exempts.

"Pre-amendments devices" refers to devices legally marketed in the US before the enactment of the Medical Device Amendments on May 28, 1976, and that has not been:

- Significantly changed or modified since then; and
- For which the FDA has not determined a Premarket Approval (PMA) application is needed to provide reasonable assurance of the device's safety and effectiveness.

For exempted devices, a Premarket Notification Application and FDA clearance are not required. However, the manufacturer is required to register its establishment and list its generic product with the FDA. It is also important to remember that all medical devices (including exempt devices) are subject to the Quality System Regulation (21 CFR 820), known as good manufacturing practices (GMP) unless there is an exception noted in 21 CFR 820.

In most cases, the medical device must be approved or cleared by the FDA to get to the market.

While words "approved" and "cleared" are often used interchanging, they are distinct categories. In the regulatory sense, the term "approved" refers to the Premarket Approval (PMA) pathway reserved for higher-risk devices. The word "cleared" refers to the so-called 510(k) clearance pathway, which is based on "substantial equivalence." Though subtle, the difference is significant as the FDA's approval implies certain protections, such as immunity from litigation in instances of harm directly related to the device, which was confirmed in the Riegel vs. Medtronic court decision.[3]

2.1.2 Premarket Notification [510(k)]

Premarket notification (or 510(k)) is the most common pathway to enter the US market. For specific Class I and most Class II devices,

---

[3]Public Citizen (2012). Substantially Unsafe: Medical Devices Pose Great Threat to Patients; Safeguards Must be Strengthened, Not Weakened. https://www.citizen.org/article/substantially-unsafe-2/

which are substantially equivalent to the existing on the market products, manufacturers can use Section 510(k) of the Food, Drug, and Cosmetic Act to notify the FDA of their intent to market a medical device. This is known as Premarket Notification (PMN) or 510(k). Under 510(k), a manufacturer must demonstrate to FDA that their device is substantially equivalent (as safe and effective) to a device already on the market, which is not subject to Premarket Approval (PMA).

While Premarket Notification sounds like a notification, it is typically a multi-stage and multi-month review process. Two primary phases of the 510(k) review are the Acceptance Review and the Substantive Review. To get the device cleared, an applicant collects data about their device's safety and efficiency and submits the 510(k) application form to the FDA. A user fee payment should accompany the submission. If the user fee payment and the valid filing have been received, the FDA issues an Acknowledgement Letter with the date and 510(k) number and proceeds to the Acceptance Review.

On clearing the Acceptance Review, a 510(k) proceeds to the Substantive Review. During Substantive Review, FDA conducts a comprehensive review of the 510(k) submission and communicates with the applicant through the so-called Substantive Interaction.

The Substantive Review results in the 510(k) Decision Letter. The Decision Letter's primary outcome determines whether the submission is substantially equivalent (SE) or not substantially equivalent (NSE).

A 510(k) that receives an SE decision is considered "cleared." FDA adds the cleared 510(k) to the 510(k) database, which is updated weekly.

2.1.2.1 Types of the 510(k) submission

There are three types of Premarket Notification 510(k) submissions: Traditional, Special, and Abbreviated. Most new devices initially go through the Traditional 510(k) pathway. The Special 510(k) and Abbreviated 510(k) submission types can be used when a 510(k) submission meets specific criteria.

Traditional 510(k)

Traditional 510(k) is the primary regulatory pathway for majority of new devices. Any significant change to the device (significant change or modification in design, components, method of manufacture, or intended use) will require submitting a new 510(k). However, some device changes may be implemented without the submission of a new 510(k).

The Traditional 510(k) target timeframes are 90 days.

Special 510(k)

Device manufacturers may choose to submit a Special 510(k) **for changes to their own existing device** if "the method(s) to evaluate the change(s) are well-established, and when the results can be sufficiently reviewed in a summary or risk analysis format."[4] For example, a Special 510(k) most likely will not be appropriate for devices that manufacture a biological product at the point of care because there would likely be no well-established method to evaluate such changes and/or the performance data would not be reviewable in a summary or risk analysis format.[5] On the other side, design or labeling changes to an existing device may be appropriate for a Special 510(k).

The FDA generally reviews Special 510(k) submissions within 30 days of receipt.

Abbreviated 510(k)

In certain circumstances, the applicant may submit an Abbreviated 510(k). The Abbreviated 510(k) filing relies on one or more:

- FDA guidance document(s), or
- Demonstration of compliance with special control(s) for the device type, or
- Voluntary consensus standard(s).

The FDA targets 90 days for the review of the Abbreviated 510(k).

---

[4]FDA, 510(k) Submission Programs. Retrieved from https://www.fda.gov/medical-devices/premarket-notification-510k/510k-submission-programs
[5]Schroeder W, (2021) Special 510(k) vs. Abbreviated 510(k) vs. Traditional 510. Retrieved from https://www.greenlight.guru/blog/special-510k-abbreviated-traditional

Note that Traditional and Abbreviated 510(k) submissions are fairly extensive documents.

Premarket notification (510(k) process) is described in more detail in Appendix A (Premarket Notification [510(k)] Process).

Important Takeaways:

- FDA targets to issue the Decision Letter within 90 calendar days since the Acknowledgement Letter (excluding any days where the submission was on hold). However, in reality, with several rounds of questions (which put the application on hold and stops a regulatory timer), this process can take much longer. The 6 to 9 months process is a more realistic scenario.

2.1.3 Premarket Approval (PMA)

Class III is reserved for devices that support or sustain human life, are of substantial importance in preventing human health impairment, or present a potential, unreasonable risk of illness or injury. Due to the level of risk associated with Class III devices, FDA has determined that general and special controls alone are insufficient to assure the safety and effectiveness of Class III devices.[6]

For Class III devices, a Premarket Approval (PMA) application is required unless the device is a "pre-amendments device" (legally marketed before May 28, 1976, or substantially equivalent to such a device) and the FDA has not determined later that PMA is required (in that case, a 510(k) will be the route to market). The applicant must receive FDA approval of its PMA application before marketing the device.

PMA is a scientific and regulatory review to evaluate the safety and effectiveness of Class III medical devices. It is the most stringent type of device marketing application required by the FDA. PMA approval is based on the FDA's determination that the submission contains sufficient valid scientific evidence to assure that the device is safe and effective for its intended use(s).

The PMA review starts with the administrative checklist. Suppose a PMA application lacks elements listed in the administrative checklist. In that case, the FDA will refuse to file a PMA application and not proceed with the in-depth review of scientific and clinical

---

[6]FDA, Premarket Approval (PMA). Retrieved from https://www.fda.gov/medical-devices/premarket-submissions/premarket-approval-pma

data. PMA applications that are incomplete, inaccurate, inconsistent, miss critical information, and poorly organized can result in delays in approval or denial of those applications.

FDA is a science-driven organization. Good science and scientific writing are vital to the approval of the PMA application. The technical sections are the most important part of the PMA submission. If a PMA application lacks valid clinical information and scientific analysis on sound scientific reasoning, it will affect review and approval.

Technical sections of the PMA are typically divided into non-clinical laboratory studies and clinical studies.

Non-clinical Laboratory Studies Section: Non-clinical laboratory studies section includes information on microbiology, toxicology, immunology, biocompatibility, stress, wear, shelf life, and other laboratory or animal tests. Non-clinical studies for safety evaluation **must be conducted** in compliance with 21 CFR Part 58 (Good Laboratory Practice for Nonclinical Laboratory Studies).[7]

Clinical Investigations Section: Clinical investigations section includes study protocols, safety, and effectiveness data, adverse reactions and complications, device failures and replacements, patient information, patient complaints, tabulations of data from all individual subjects, results of statistical analyses, and any other information from the clinical investigations.[8] For example, the submission must include Form FDA-3674, the Certification of Compliance with the Requirements of ClinicalTrials.gov Data Bank. Any study conducted under an Investigational Device Exemption (IDE) must be identified as such.

In addition to three primary pathways, there are several additional ones:

- Humanitarian Device Exemption (HDE): can be used for devices, which aim to treat or diagnose a disease or condition that affects fewer than 8,000 individuals in the United States per year

---

[7]FDA, (2019) Recommended Content and Format of Non-Clinical Bench Performance Testing Information in Premarket Submissions. Retrieved from https://www.fda.gov/regulatory-information/search-fda-guidance-documents/recommended-content-and-format-non-clinical-bench-performance-testing-information-premarket
[8]LMG, Premarket Approval (PMA). Retrieved from https://www.fdahelp.us/premarket-approval.html

- De Novo Classification: reserved for novel devices with low and medium risks
- Breakthrough Devices: reserved for novel devices with high impact on patients

2.1.4   Humanitarian Device Exemption (HDE)

HDE provides a regulatory pathway for Class III devices intended to benefit patients with rare diseases or conditions.

A Humanitarian Use Device (HUD) is a device that is intended to benefit patients by treating or diagnosing a disease or condition that affects fewer than 8,000 individuals in the United States per year.[9] Such as research and development costs could exceed its market returns for diseases or conditions affecting small patient populations. FDA introduced HUD regulation in 1990 to provide incentives for developing devices for use in the treatment or diagnosis of diseases affecting these populations.

The Humanitarian Device Exemption (HDE) application is similar in both form and content to a Premarket Approval (PMA) application but is exempt from the effectiveness requirements of a PMA. In particular, an HDE application is not required to contain the results of scientifically valid clinical investigations demonstrating that the device is effective for its intended purpose.[10] However, it must have sufficient information for the FDA to determine that the device does not pose an unreasonable or significant risk of illness or injury and that the probable benefit to health outweighs the risk of injury or illness from its use. Additionally, "the applicant must demonstrate that no comparable devices are available to treat or diagnose the disease or condition and that they could not otherwise bring the device to market."[11]

An approved HDE authorizes the marketing of the device. However, a HUD device may only be used after IRB approval has been obtained to use the device for the FDA-approved indication.

---

[9]FDA, Humanitarian Use Devices and Humanitarian Device Exemption. Retrieved from https://www.fda.gov/ScienceResearch/SpecialTopics/ PediatricTherapeuticsResearch/ucm249423.htm.

[10]FDA, Information Sheet Guidance for IRBs, Clinical. Retrieved from https://www. fda.gov/media/75381/download

[11]FDA, HDE Approvals, Retrieved from. https://www.fda.gov/medical-devices/ device-approvals-denials-and-clearances/hde-approvals

2.1.5 De Novo Classification

The FDA automatically considers any completely new (novel) device as a high-risk (Class III) device, which has very high costs associated with its approval. To lower the regulatory costs for low and medium-risk devices, in 1997 FDA added the De Novo classification option as an alternate pathway to classify novel medical devices.

Initially, De Novo was designed as an option for devices that received a "not substantially equivalent" (NSE) determination in response to a Premarket Notification [510(k)] submission. In 2012 the legislation was amended to allow an applicant to submit a De Novo classification request to the FDA without submitting a 510(k).

Devices that are classified through the De Novo process may be marketed. Note that De Novo's approval creates a new product code.

As of the time of writing, there are two options for De Novo classification for low to moderate risk novel devises:[12]

- "Any person who receives an NSE determination in response to a 510(k) submission may, within 30 days of receipt of the NSE determination, submit a De Novo to Class I or II, or
- Any person who determines that there is no legally marketed device upon which to base a determination of substantial equivalence may submit a De Novo request for the FDA to make a risk-based classification of the device into Class I or II without first submitting a 510(k)."

2.1.6 Breakthrough Devices

FDA implemented the Breakthrough Devices Program in 2018.[13] It replaced the Expedited Access Program that was introduced in 2015. The program gained popularity with 11 designations in 2016, 19 in 2017, 55 in 2018, and 50 designations as of May 2020. For example, Elon Musk's Neurolink project got a Breakthrough Device designation in July 2020.

---

[12]FDA, Evaluation of Automatic Class III Designation (De Novo). Retrieved from https://www.fda.gov/about-fda/cdrh-transparency/evaluation-automatic-class-iii-designation-de-novo-summaries

[13]FDA, Breakthrough Devices Program: Guidance document. Retrieved from https://www.fda.gov/media/108135/download

The guidance states that[14] "devices subject to PMAs, a Premarket Notification (510(k)), or requests for De Novo designation are eligible for Breakthrough Device designation if both of the following criteria are met:

(i) The device provides for more effective treatment or diagnosis of life-threatening or irreversibly debilitating human disease or conditions

(ii) The device also meets at least one of the following:

    a. Represents breakthrough technology

    b. No approved or cleared alternatives exist

    c. Offers significant advantages over existing approved or cleared alternative

    d. Device availability is in the best interest of patients"

Out of these two criteria, criterion (i) is of primary importance. The new device needs to be compared with the standard of care, which is (a) received marketing authorization for the indication being considered, and (b) currently marketed in the US and is a relevant option for patients with the identified disease or condition.

Breakthrough Device designation request enables supplementary interaction mechanisms with FDA in addition to the regular pre-submission. It also facilitates the Data Development Plan (DDP) and Sprint mechanisms, which allow for accelerating communications with the FDA 10× (7 days instead of 70 days for pre-submission).

As a result, the Breakthrough Device designation can significantly decrease time to market, particularly for devices, which require Premarket Approval. The decision to grant or deny Breakthrough Device designation is typically issued within 60 calendar days. In addition to that, a breakthrough designated device may receive Premarket Approval (PMA) without completion of a pivotal clinical trial. Namely, the FDA may allow using timely post-market data collection for breakthrough designated device subject to PMA when scientifically justified. However, the device must still meet the statutory standard of reasonable assurance of safety and effectiveness at the time of approval.

---

[14]FDA, Breakthrough Devices Program: Guidance document. Retrieved from https://www.fda.gov/media/108135/download

In some circumstances (if the manufacturer has a good track record for quality management compliance), the FDA may accept less quality management and manufacturing information in a PMA. In this case, the manufacturer must satisfy the statutory and regulatory requirements using an alternative approach to submitting all the items listed in FDA guidance.[15]

The Breakthrough Device designation will also allow using an expedited way through the reimbursement system, which will be addressed in Chapter 8.

2.1.7 Medical Device Classification

So, as you see, the US regulatory landscape is quite complicated. How to navigate through it and determine the class of your medical device? In other words, how to classify the medical device? There are several approaches to address this question.

Classification of the device relies on several concepts. Firstly, the device classification depends on the intended use of the device and indications for use. Secondly, the classification is risk-based. Namely, the risk, which the device poses to the patient and/or the user is a major factor in the class determination. Class I is assigned to devices with the lowest risk, and Class III includes those with the greatest risk.

Most medical devices can be classified by finding the matching description of the device in Title 21 of the Code of Federal Regulations (CFR), Parts 862–892. Thus, to find the device's classification (and whether any exemptions exist), you need to find the regulation number (classification regulation) for the device. It can be done either through a) the classification database search for the device name or b) the device panel (medical specialty) to which the device belongs.

FDA maintains multiple databases. The most helpful source for the classification of the device is a Product Code Classification Database.[16] The classification database will provide the classification

---

[15]FDA, Quality System Information for Certain Premarket Application Reviews: Guidance document. Retrieved from https://www.fda.gov/regulatory-information/search-fda-guidance-documents/quality-system-information-certain-premarket-application-reviews

[16]FDA, Product Classification database. Retrieved from https://www.accessdata.fda.gov/scripts/cdrh/cfdocs/cfPCD/classification.cfm

panel (e.g., orthopedic devices), common name, product code, and CFR regulation if the device type has received final classification by FDA (e.g., 21 CFR 888.1100, arthroscope).

You can use a part of the device name or a keyword search through the quick search option in the classification database. In most cases, the database will identify the classification regulation in the CFR. The goal of such exercise is to find a seven-digit number of the appropriate regulation (e.g., for a clinical mercury thermometer it will be 21 CFR 880.2920 - Clinical Mercury Thermometer), the class (Class II in this example), and a three-letter product code (e.g., FLK for clinical mercury thermometer).

Alternatively, you can perform the search through the panel. FDA has classified and described over 1,700 distinct types of devices and organized them into 16 medical specialties or "panels" (anesthesiology, cardiovascular, chemistry, dental, ear, nose, and throat, gastroenterology and urology, general and plastic surgery, general hospital, hematology, immunology, microbiology, neurology, obstetrical and gynecological, ophthalmic, orthopedic, pathology, physical medicine, radiology, and toxicology). So, you can go directly to an appropriate panel and perform a search there.

Search through the FDA website and databases will help you identify the appropriate guidance documents, recognized consensus standards, and find potential predicate devices.

In well-established fields, FDA developed guidance documents to help industry and FDA staff to streamline the approval process. For example, for pulse oximeter [codes: DQA (oximeter), NLF (oximeter, reprocessed), DPZ (ear oximeter)] FDA developed already several generations of guidance documents.[17]

Another important outcome of the classification database search is a product code. The product code is particularly useful for searching a predicate device, which is an essential element for establishing a "substantial equivalence" (SE) required in Premarket Notification, or 510(k) process.

---

[17]FDA, (2013) Pulse Oximeters - Premarket Notification Submissions [510(k)s]: Guidance for Industry and Food and Drug Administration Staff. Retrieved from https://www.fda.gov/regulatory-information/search-fda-guidance-documents/pulse-oximeters-premarket-notification-submissions-510ks-guidance-industry-and-food-and-drug

2.1.8   Predicate Device

As we mentioned before, Premarket Notification [510(k)] is a premarketing submission made to FDA to demonstrate that the device to be marketed is safe and effective by proving substantial equivalence (SE) to a legally marketed device.[18] To demonstrate SE, the applicant must compare their device to a similar, legally marketed US device(s). The legally marketed device to which equivalence is drawn is termed the predicate device.

"Substantial equivalence" claim can be made against a variety of devices. It can be any before 1976 or post-1976 device that **is** or **was** legally marketed in the United States. For example, an applicant may claim SE to a device that is no longer being marketed in the United States. However, SE cannot be claimed against a device found in violation of the Federal Food Drug & Cosmetic (FD&C) Act.

A claim of substantial equivalence does not mean the device must be identical. Substantial equivalence is based on the intended use. It can be established for design, energy used or delivered, materials, performance, safety, effectiveness, labeling, biocompatibility, standards, and other applicable characteristics.

Three critical elements for establishing substantial equivalence are (see Appendix A for further details): (a) the predicate device is or was legally marketed in the United States, (b) the same intended use, and (c) technological characteristics. If technical characteristics are the same as a predicate device, it is a good base for establishing SE. If they are different (e.g., you are using a different physical principle), then you need to prove that it is safe and efficient, validation methods are adequate, and performance data demonstrate substantial equivalence.

While you can use any device, which is or was legally marketed in the United States, in most cases, a device recently cleared through 510(k) is used as a predicate device.

The FDA databases typically contain the summary of each cleared 510(k) device. It includes information on the indications for use, the predicate device, the device description, comparison with the predicate device, and a summary of performance testing (e.g.,

---

[18]FDA, How to Find and Effectively Use Predicate Devices. Retrieved from https://www.fda.gov/medical-devices/premarket-notification-510k/how-find-and-effectively-use-predicate-devices

standards, which were tested against). The exception is made for 510(k) statement submission (see Appendix A for details). In this case, there are almost no details publicly available for such a device. It is probably easier to use the device filed through a summary rather than a statement as a predicate device.

2.1.9 513(g) Request

If you are unsure about the device's classification, you can get informal or formal assistance from the FDA.

For informal assistance, you can either contact the Division of Industry and Consumer Education (DICE) or the Device Determination experts (DeviceDetermination@fda.hhs.gov). Obviously, their responses are not binding. Namely, they are not classification decisions and do not constitute FDA clearance or approval for commercial distribution.

If you want a formal device determination from the FDA, you can submit a 513(g) Request for Information.[19] According to Section 513(g) of the FD&C Act (21 USC 360c(g)), "Within sixty days of the receipt of a written request of any person for information respecting the class in which a device has been classified or the requirements applicable to a device under this act, the secretary shall provide such person a written statement of the classification (if any) of such device and the requirements of this act applicable to the device." Typically, the FDA response contains the following information:[20]

- "the FDA assessment as to the generic type of device (e.g., classification regulation) if any;
- the class of devices within that generic type;
- whether a PMA, 510(k), or neither is required in order to market devices of the particular class within that generic type;
- other requirements applicable to devices of the particular class within that generic type;

---

[19]FDA (2019), FDA and Industry Procedures for Section 513(g) Requests for Information under the Federal Food, Drug, and Cosmetic Act. Retrieved from https://www.fda.gov/media/78456/download
[20]FDA (2019), FDA and Industry Procedures for Section 513(g) Requests for Information under the Federal Food, Drug, and Cosmetic Act. Retrieved from https://www.fda.gov/media/78456/download

- whether a guidance document has been issued regarding the exercise of enforcement discretion over the particular class of devices within that generic type;
- whether additional FDA requirements may apply (e.g., for radiation-emitting products)."

In response to the 513(g) request, FDA will not provide an opinion about "substantial equivalence," will not review information about safety and effectiveness, and will not address questions regarding the specific types of testing (non-clinical, animal, or clinical) appropriate to support clearance or approval of the device. FDA's responses to 513(g) requests for Information are not device classification decisions and cannot be considered an FDA clearance or approval for marketing. Technically, a 513(g) response does not constitute a final FDA decision about classification; however, it is a good starting point for premarket submissions for approval or clearance.

2.1.10 Special Cases

In addition to general guidance documents and device-specific standards, FDA has developed several special cases and approaches that may deviate from the standard definitions and traditional pathways. Below, we will consider several such medical device types and approaches that may have a particular relevance nowadays.

2.1.10.1 Digital health

Digital health has a broad scope and includes mobile health (mHealth), health information technology (IT), wearable devices, telehealth and telemedicine, and personalized medicine.

2.1.10.2 Software as a medical device

Certain software can be considered as a medical device and regulated. In this case, it is referred to as software as a medical device (SaMD). SaMD can range from an app to dose-painting software in cancer treatment.

How to determine whether an app or a system for data collection from remote sensors is a medical device or not? And how will the FDA interpret it? SaMD regulation is a good example of international cooperation to achieve harmonization between national approaches. In its SaMD regulations, the FDA adopted an approach based on the International Medical Device Regulators Forum's (IMDRF) recommendations. According to IMDRF, software as a medical device is "software intended to be used for one or more medical purposes that perform these purposes without being part of a hardware medical device."

An essential part of this definition is that if the software is part of a hardware medical device, it does not meet the definition of software as a medical device.[21] So, it is not necessary to file separate filings for the hardware and firmware.

IMDRF adopted a category-based approach. The four categories (I, II, III, and IV) are based on the levels of impact on the patient or public health where accurate information provided by the SaMD to treat or diagnose, drive or inform clinical management is vital to avoid death, long-term disability, or other serious deterioration of health, mitigating public health (see Table 2.1. for details). The categories are of relative significance to each other. Category IV has the highest level of impact, Category I the lowest. We have included examples of software, which corresponds to various cells of this table in Appendix B.

Table 2.1 SaMD Categories

State of healthcare situation or condition	Significance of information provided by SaMD to healthcare decision		
	Treat or diagnose	Drive clinical management	Inform clinical management
Critical	IV	III	II
Serious	III	II	I
Non-serious	II	I	I

Source: Adapted from Ref. [22] with permission.

---

[21]FDA, What are examples of Software as a Medical Device? Retrieved from https://www.fda.gov/medical-devices/software-medical-device-samd/what-are-examples-software-medical-device

[22]IMDRF (2014), "Software as a Medical Device:" Possible Framework for Risk Categorization and Corresponding Considerations. Retrieved from http://www.imdrf.org/docs/imdrf/final/technical/imdrf-tech-140918-samd-framework-risk-categorization-141013.pdf

Note that SaMD categories are better aligned with the European and Canadian medical device classification (will be discussed later). In the case of the United States, four categories require some additional adjustment to the three-class system.

Important Takeaways:

- In the case of the United States market, SaMD is treated as any other medical device. It requires a regular premarket submission [510(k) or PMA] unless it is exempt.
- Due to the complexity of regulations, it is wise to understand whether the software is a medical device early in the development process. Reviewing FDA guidances is an excellent way to start this process. These guidances can help determine whether the software is a medical device or not and understand what kind of evidence the FDA expects to see in the premarket submission. In particular, for clinical evidence, one can use the Guidance for Software as a Medical Device (SaMD): Clinical Evaluation,[23] which is based on the IMDRF document.

2.1.10.3 Mobile medical apps

Due to widespread digitalization in healthcare, in addition to software as a medical device (SaMD), the FDA has a concept of a mobile medical apps (MMAs). These concepts started merging recently;[24] however, due to a significant number of apps, which can fall under the definition of a medical device, this topic is worth considering in some level of detail.

Mobile medical apps are mobile apps that meet a medical device's definition and are an accessory to a regulated medical device or transform a mobile platform into a regulated medical device.[25] More specifically, an MMA is intended for any of the following:

- "use as an accessory to a regulated medical device (for example, an app that alters the function or settings of an infusion pump)

---

[23]FDA (2017), Software as a Medical Device (SAMD): Clinical Evaluation. Retrieved from https://www.fda.gov/media/100714/download
[24]FDA (2019), Policy for Device Software Functions and Mobile Medical Applications. Retrieved from https://www.fda.gov/regulatory-information/search-fda-guidance-documents/policy-device-software-functions-and-mobile-medical-applications
[25]FDA, Device Software Functions Including Mobile Medical Applications. Retrieved from https://www.fda.gov/medical-devices/digital-health-center-excellence/device-software-functions-including-mobile-medical-applications

- transforming a mobile platform into a regulated medical device (for example, an app that uses an attachment to the mobile platform to measure blood glucose levels)
- performing sophisticated analysis or interpreting data from another medical device (for example, an app that uses consumer-specific parameters and creates a dosage plan for radiation therapy)"[26]

FDA developed a guidance document,[27] where they split mobile medical apps into three buckets and provided examples of software functions for each of them. These buckets include healthcare-related mobile apps that:

- are not medical devices,
- are medical devices, but for which the FDA intends to exercise enforcement discretion, and
- are medical devices and are the focus of FDA oversight.

The FDA focuses its regulatory oversight on a small subset of health apps that pose a higher risk if they don't work as intended. For many software functions that meet the regulatory definition of a "medical device" but pose minimal risk to patients and consumers, the FDA exercises enforcement discretion and does not expect manufacturers to submit premarket review applications or register and list their software FDA. This includes device software that:[28]

- "Help patients/users self-manage their disease or condition without providing specific treatment suggestions; or
- Automate simple tasks for healthcare providers."

The FDA issued the first MMA guidance in 2013. Since then, the FDA revised the guidance several times and clarified that software policies are function-specific and apply across platforms. Based

---

[26]Federal Trade Commission, Mobile Health Apps Interactive Tool. Retrieved from https://www.ftc.gov/tips-advice/business-center/guidance/mobile-health-apps-interactive-tool

[27]FDA (2019), Policy for Device Software Functions and Mobile Medical Applications. Retrieved from https://www.fda.gov/regulatory-information/search-fda-guidance-documents/policy-device-software-functions-and-mobile-medical-applications

[28]FDA, Device Software Functions Including Mobile Medical Applications. Retrieved from https://www.fda.gov/medical-devices/digital-health-center-excellence/device-software-functions-including-mobile-medical-applications

on that determination, the FDA changed definitions from "mobile application" to "software function" recently.

Software functions excluded from the device definition include:

- administrative support for healthcare facilities
- encouragement of healthy lifestyles
- electronic patient records
- transferring, storing, converting formats, or displaying data
- providing limited clinical decision support

There are several important takeaways about MMAs:

- While the FDA applies risk-based approaches to medical devices, certain MMAs can be exempt from regulations. Even if they meet the medical device's definition, the FDA may exercise enforcement discretion and does not expect manufacturers to submit premarket review applications or register and list their software with the FDA.
- FDA understands the iterative nature of software development. Thus, the FDA's device software functions and mobile medical apps policies do not require software developers to submit for FDA clearance/approval minor, iterative product changes.
- The FDA's mobile medical apps policy does not consider mobile platform manufacturers to be medical device manufacturers just because their mobile platform could be used to run a mobile medical app regulated by the FDA.[29]

2.1.10.4 Artificial intelligence and machine learning as a medical device

Adaptive artificial intelligence (AI) and machine learning (ML) technologies do not fit well into the FDA's traditional risk-based Medical Device Regulation paradigm. The training or retraining models based on the new data under the FDA's current approach to software modifications will fall under the definition of modifications that may require a premarket review, which completely defies the purpose.

---

[29]FDA, Device Software Functions Including Mobile Medical Applications. Retrieved from https://www.fda.gov/medical-devices/digital-health-center-excellence/device-software-functions-including-mobile-medical-applications

Understanding the potential of these technologies, FDA considers a total product lifecycle-based regulatory framework for AI technologies. This framework would allow modifications to be made from real-world learning and adaptation while still ensuring that safety and effectiveness are maintained.

FDA has not issued a guidance document yet. To initiate the process, FDA solicited public opinion by issuing a Discussion Paper and Request for Feedback.[30]

In the proposed framework, the FDA introduced a "predetermined change control plan" in premarket submissions. This plan would include the types of anticipated modifications ("Software as a Medical Device Pre-Specifications") and the associated methodology being used to implement those changes in a controlled manner that manages risks to patients ("Algorithm Change Protocol.")

In this approach, the FDA would expect a commitment from manufacturers on transparency and real-world performance monitoring for artificial intelligence and machine learning-based software as a medical device and periodic updates to the FDA on what changes were implemented as part of the approved pre-specifications and the algorithm change protocol.[31]

2.1.10.5   Medical device accessories

A medical device may be used in conjunction with other devices. Suppose an infusion pump system includes an infusion pump and a stand. The stand holds medications and liquids at an appropriate height and supports the infusion pump's performance. Based on its intended use, that secondary device (stand) can be considered a medical device accessory and regulated. According to the FDA, an accessory "is a finished device that is intended to support, supplement, and/or augment the performance of one or more parent devices."[32] If labeling, promotional materials, or other evidence of intended use demonstrates that the device is intended to support,

---

[30]FDA, Proposed Regulatory Framework for Modifications to Artificial Intelligence/ Machine Learning (AI/ML)-Based Software as a Medical Device (SaMD). Retrieved from https://www.fda.gov/media/122535/download

[31]FDA, Artificial Intelligence and Machine Learning in Software. Retrieved from https://www.fda.gov/medical-devices/software-medical-device-samd/artificial-intelligence-and-machine-learning-software-medical-device

[32]FDA, Medical Device Accessories. Retrieved from https://www.fda.gov/medical-devices/classify-your-medical-device/medical-device-accessories

supplement, and/or augment another device, whether a particular brand or a device type, that device is considered an accessory. In our example, the infusion pump would be considered the parent device, and the stand would be viewed as the accessory to the infusion pump.

Using something in conjunction with a medical device does not necessarily make that something an accessory. It depends on the intended use. For example, a mobile smartphone would not be considered an accessory after downloading a medical application (app) because it was not specifically intended for use with the medical device.

To be considered an accessory, the article needs (a) to be intended for use with a parent device(s) and (b) to be intended support, supplement, and/or augment the performance of a parent device(s)

Traditionally, the accessory could be classified either (a) with the parent device or (b) by issuance of a unique, separate classification regulation for the accessory. Since 2017, with amendments to the FD&C Act, the FDA classifies accessories based on the accessory risks when used as intended. Thus, the accessory classification does not depend on the classification of the parent device. For example, an accessory to a Class III parent device may pose a lower risk that could be mitigated through general controls or general and special controls and could be regulated as Class I or Class II, respectively.[33]

An accessory can be classified using an Accessory Classification Request,[34] a written request submitted to the FDA under Section 513(f)(6) of the FD&C Act. There are two types of Accessory Classification Requests:

- A New Accessory Request is appropriate for an accessory of a type that has not been previously classified under the FD&C Act, cleared for marketing under a 510(k) submission, or approved in a PMA.
- An Existing Accessory Request is appropriate for an accessory that has been granted marketing authorization as part of a submission for another device.

---

[33]FDA, Medical Device Accessories. Retrieved from https://www.fda.gov/medical-devices/classify-your-medical-device/medical-device-accessories
[34]FDA (2017), Medical Device Accessories: Describing Accessories and Classification Pathways. Retrieved from https://www.fda.gov/media/90647/download

An Existing Accessory Classification Request is a standalone request (i.e., not included as part of a PMA or 510(k) application) from an applicant who already has marketing authorization for their accessory seeking proper classification of an existing accessory type.[35]

A New Accessory Classification Request is a request included in a PMA or 510(k) submission to classify a new accessory type properly.

For new accessory types, in addition to the Accessory Requests, the applicant can classify the accessory through the De Novo classification process (Section 513(f)(2) of the FD&C Act).

2.1.10.6 Pre-Cert program

To streamline regulatory oversight in the SaMD space, the FDA developed the Software Pre-certification (Pre-Cert) Program. As of the time of writing, it is in the pilot stage. The program aims to develop a regulatory model that will provide more streamlined and efficient regulatory oversight of software-based medical devices. The program targets manufacturers who have demonstrated a robust culture of quality and organizational excellence and are committed to monitoring their products' real-world performance once they reach the US market.

This program shifts oversight focus from the product (currently performed for traditional medical devices) to the software developer or digital health technology developer.

The process starts with the Excellence Appraisal, where the FDA assesses capacities, capabilities, and controls. The FDA mainly focuses on five core areas: patient safety, product quality, clinical responsibility, cybersecurity responsibility, and proactive culture. Based on the assessment results, the FDA performs Review Determination and Streamlined Review of the product (if required).

Potentially, pre-certified companies could market their lower-risk devices without the FDA's premarket review or only a streamlined premarket review based on its pre-certification level and IMDRF risk categorization.

Finally, after the product has reached the market, the company gathers real-world performance, which will be the input for future Excellence Assessments.

---

[35]FDA, Medical Device Accessories. Retrieved from https://www.fda.gov/medical-devices/classify-your-medical-device/medical-device-accessories

In 2017 FDA selected nine companies (Apple, Fitbit, Johnson & Johnson, and Roche, among others) to conduct the pilot. Based on the results of the pilot, the program most likely will be amended. However, hopefully, it will preserve the initial aims.

2.1.10.7 Emergency use authorization

In certain circumstances, medical devices can be brought to the market using nonstandard mechanisms. The United States has an extensive fabric of emergency preparedness and response legislation. In particular, FDA has the Medical Countermeasures Initiative (MCMi). Under Section 564 of the FD&C Act, the FDA commissioner may allow unapproved medical products or unapproved uses of approved medical products to be used in an emergency to diagnose, treat, or prevent serious or life-threatening diseases or conditions caused by chemical, biological, radiological, and nuclear (CBRN) threats when there are no adequate, approved, and available alternatives. It is known as an Emergency Use Authorization (EUA).

For example, in 2020, the FDA issued several EUAs for coronavirus disease 2019 (COVID-19). Before that, the FDA issued EUAs for the Zika virus (2016), Enterovirus virus D68 (EV-D68) (2015), Ebola virus (2014), coronavirus (2013), and H7N9 influenza (2013).

The EUA mechanism allows entering the US market faster and with less scrutiny. However, it does not mean there is no oversight. In particular, FDA issued the EUA guidance document.[36] For example, the FDA still expects safety and performance testing to be performed. However, full test results need to be submitted not upfront but rather parallel with the approval process. The manufacturer still is a subject of post-market regulations (e.g., report any adverse effects). Also, it should be noted that EUA is temporary permission. When an emergency is declared over, the FDA revokes the authorization, and the device cannot be legally marketed anymore. Thus, it is important to apply for market entry via a regular premarket submission [510(k), PMA, or De Novo].

---

[36]FDA (2017), Emergency Use Authorization of Medical Products and Related Authorities: Guidance for Industry and Other Stakeholders. Retrieved from https://www.fda.gov/media/97321/download

2.1.11   Closing Remarks

As one can see, the regulatory requirements to enter the US market are quite extensive. However, we just scratched the regulatory surface. So far, we discussed only the premarket submission (if not exempt) requirements. In addition to them, the basic regulatory requirements that manufacturers of medical devices distributed in the United States must comply with are:

- Establishment registration,
- Medical device listing,
- Quality system (QS) regulation,
- Labeling requirements, and
- Medical device reporting (MDR).

Establishment registration—manufacturers (domestic and foreign) and distributors/importers of medical devices must register their establishments with the FDA and pay fees annually. In addition to registration, foreign manufacturers must also designate a US Agent.

Medical Device Listing—Manufacturers must list their devices with the FDA. Establishments required to list their devices include, among others are manufacturers, contract manufacturers, contract sterilizers that commercially distribute the device, re-packagers, and re-labelers, specification developers, re-processors single-use devices, remanufacturers, manufacturers of accessories and components sold directly to the end-user, US manufacturers of "export only" devices.

Quality system regulation (QS regulation)—The QS regulation includes requirements related to the methods used and the facilities and controls used to design, purchase, manufacture, packaging, label, store, install, and servicing medical devices. Manufacturing facilities undergo periodical FDA inspections to assure compliance with the QS requirements. The manufacturer needs to design and implement a quality management system (QMS) before premarket submission.

Labeling—labeling includes labels on the device and descriptive and informational literature that accompanies the device.

Medical device reporting (MDR)—Incidents in which a device may have caused or contributed to a death or serious injury must be reported to FDA under the Medical Device Reporting Program. Besides, certain malfunctions must also be reported. The MDR regulation is a mechanism for the FDA and manufacturers to identify and monitor significant medical devices' significant adverse events. The goals of the regulation are to detect and correct problems in a timely manner.

In addition to that, there are multiple other non-FDA regulations, which impact medical devices:

- Health Insurance Portability and Accountability Act (HIPAA). HIPAA rules protect certain health information's privacy and security and require certain entities to notify health information breaches.
- Federal Trade Commission Act (FTC Act). The FTC enforces the FTC Act, which prohibits deceptive or unfair acts or practices affecting commerce, including privacy and data security and those involving false or misleading claims about apps' safety or performance.
- FTC's Health Breach Notification Rule. The FTC's Health Breach Notification Rule requires certain businesses to provide notifications following personal health record information breaches.[37]

Important Takeaways:

- Manufacturers of medical devices are subjects of multiple regulations. Start getting familiar with these regulations early in the project.
- The most significant regulations that impact the project's complexity and timelines are premarket submission and Quality System Regulation.
- You need to set reasonable expectations about premarket submissions. For example, for 510(k), while the FDA targets 90–100 calendar days for approval, you can expect 1 or 2 rounds of questions in a realistic case, which stops the regulatory timer. Thus, a 6 to 9 months timeframe is much more real.

---

[37]Federal Trade Commission, Mobile Health Apps Interactive Tool. Retrieved from https://www.ftc.gov/tips-advice/business-center/guidance/mobile-health-apps-interactive-tool

2.2   European Union

EU developed a regulatory approach, which is significantly different from the US one. Firstly, it is a much more streamlined rule-based process without multiple exceptions. To some extent, this difference is similar to legal systems in the respective countries: civil law in continental Europe and common law in English-speaking countries. Thus, while the US regulatory system relies on precedents (predicate devices), the EU approach relies on well-defined classification. So, classification can be easily derived from the core documents (directives in this case).

The second distinct feature of the EU MedDev market is the absence of a centralized body (e.g., FDA), which reviews applications and approves or clears submissions. Instead, the EU relies on notified bodies' decentralized network, which are third parties accredited by European authorities to audit medical device companies and products. Notified bodies evaluate that the product is safe and functions as described, but not that the technology is effective or provides a clinical benefit to the patient.[38]

As of the time of writing, medical device directives [93/42/EEC—Medical Devices Directive (MDD) and 90/385/EEC—Active Implantable Medical Devices Directive (AIMDD)] are the core of existing EU regulations.

Similar to the United States, medical devices in the EU are split into three classes (Class I, II, and III) based on their risk profiles. However, Class I and Class II devices are further divided into subclasses. Thus, the whole EU classification consists of six buckets:

(1) Class I non-sterile, non-measuring function—Low-risk devices such as bandages, compression hosiery, or walking aids

(2) Class I sterile—Low-risk devices, which require sterility. They include stethoscopes, examination gloves, colostomy bags, and oxygen masks

(3) Class I measuring function—Low-risk devices with measurement function. Examples include thermometers and non-invasive blood pressure measuring devices

---

[38]Sorenson C, Drummond M (2014). Improving medical device regulation: The United States and Europe in perspective. *Milbank Quarterly* 92, No. 1:114–150.

(4) Class IIa—Low- to medium-risk devices installed within the body in the short term. Class IIa devices are those which are installed within the body for only between 60 minutes and 30 days. Examples include hearing aids, blood transfusion tubes, and catheters

(5) Class IIb—Medium- to high-risk devices installed within the body for periods of 30 days or longer. Examples include ventilators and intensive care monitoring equipment

(6) Class III—High-risk devices. Examples include balloon catheters, prosthetic heart valves, pacemakers, etc.

The classification of the device can be derived using the Annex IX of the MDD. It consists of a set of rules, which give an unambiguous determination of the Class. Active implantable medical devices are typically subject to the same regulatory requirements as Class III devices.

The certification process aims to get a so-called CE Marking for the device and ISO 13485 certification for the facility. The manufacturer needs to implement a quality management system (QMS) and prepare a technical file for the device (a Design Dossier for Class III devices).

The next step is to appoint an Authorized Representative (EC REP) located in the EU who is qualified to handle regulatory issues.

After that, you can go to a notified body, which will review documents and audit processes and facilities.

After successfully completing the notified body audit, the device will receive a CE Marking certificate, and the facility will receive an ISO 13485 certificate. ISO 13485 certification must be renewed every year (so, you will be audited by a notified body every year). CE Marking certificates are typically valid for three years but are generally reviewed annually at the same time as the ISO 13485 surveillance audit. Failure to pass the audit will invalidate the CE Marking certificate.

The certification process's final step is to prepare a Declaration of Conformity, a legally binding document prepared by the manufacturer stating that the device complies with the applicable directive. After that, the manufacturer may affix the CE Marking and start marketing the device.

Class I non-sterile devices with non-measuring functions are exempt from the QMS requirements and notified body audits. Instead, they use a self-certified CE Marking. The self-certified CE Marking certificate does not expire as long as the manufacturer complies with the MDD.

The overall process takes 3–5 months (less for Class I non-sterile, non-measuring, slightly more for Class III).

However, this well-established system is changing. EU is moving toward CE Marking certification using European Commission Regulation (EU) No. 2017/745, commonly known as the Medical Device Regulation (MDR).

The MDR system retains all significant features of the MDD. However, it introduces much more rigorous requirements for device manufactures. The key differences of the MDR over MDD are:

- Introduction of Class I reusable surgical devices
- Class I self-certified device manufacturers are now required to implement QMS. However, they still do not require a notified body audit
- More rigorous requirements for QMS. Namely, QMS must include clinical evaluation, post-market surveillance (PMS), and post-market clinical follow-up (PMCF) plans
- Registration of the device and its unique device identifier (UDI) in the EUDAMED database
- CE Marking certificates are valid for a maximum of 5 years

Under MDR, the device's classification can be determined using Annex VIII (Classification Criteria).

A 3-year MDR transition period started in May 2017. The initial timelines for the MDR implementation are:

May 26, 2020—completion deadline for the MDR

Thus, the original implementation was scheduled for May 26, 2020. However, this deadline was pushed by one year due to the COVID-19 pandemic. The EUDAMED database implementation was delayed by two years in December 2019.

Thus, as of the time of writing, the timelines for the MDR implementation are:

May 26, 2021—full application for the MDR

May 26, 2022—full application for IVDR (*in vitro* diagnostic regulation)

May 2024—Available for all EC certificates issued five years from the date of issue/renewal, or three years from the application date of the MDR (May 27, 2021), whichever comes first

May 2025—Can no longer sell or distribute devices certified under the MDD

Another difference in the EU regulatory approach from the United States is that *in vitro* diagnostic devices (IVD) are regulated separately from medical devices (MDR). In particular, IVD devices have a classification (Class A, B, C, and D) different from MedDev classification (Class I, II, and III).

Currently, IVD devices are regulated by *In Vitro* Diagnostic Device Directive (IVDD). *In vitro* diagnostic devices will soon be regulated by *In Vitro* Diagnostic Regulations (IVDR), which have an implementation schedule similar to MDR. It is expected to be rolled out in May 2022. In May 2024, existing IVDD CE certificates will expire, and by May 2025, all remaining IVDD CE devices should be pulled off the EU market.

Key Takeaways:

- The EU MedDev certification system is very different from the United States: (a) it is rule-based, and (b) decentralized. It relies on a network of notified bodies, third-party organizations for review and audit.
- To get to the EU market, the device needs to have a CE Marking certificate, and the manufacturer needs to receive an ISO 13485 certificate.
- A manufacturer is subject to annuals reviews by a notified body. Failure to pass the audit will invalidate the CE Marking certificate.
- The EU transitions from the Medical Devices Directive (MDD) to the Medical Device Regulation (MDR) certification system. Its implementation is scheduled for May 2021.

2.3   Canada

Canada has a regulatory system, which contains elements of both centralized and decentralized systems.

Health Canada regulates the Canadian medical device market. The Canadian medical device classification is risk-based. There are four classes of medical devices: I–IV. Similar to the EU, the device classification is rule-based. It is currently regulated by the Canadian Medical Devices Regulations (CMDR) SOR/98-282. Classes I, II, III, and IV in Canada correspond to classes I, IIa, IIb, and III in the EU, respectively.

Similar to the EU, to start the regulatory approval process for Class II–IV device manufacturers, the ISO 13485 compliant quality management system (QMS) must be implemented and audited. The Canadian version of ISO 13485 standard contains certain additional requirements. So, a standard ISO 13485 certification, often used by companies selling in the European market, does not automatically meet Canadian CMDR.

Class I device manufacturers are not required to get an ISO-compliant quality management system implemented and audited. However, the manufacturer needs to apply for Medical Device Establishment License (MDEL) and implement mandatory procedures for distribution of records, complaint handling, and mandatory problem reporting.

The first step of the regulatory pathway for Class II–IV device manufacturers will be to have the quality management system audited by a registrar accredited by Health Canada under the Canadian Medical Devices Conformity Assessment System (CMDCAS). The good news for European manufacturers is that Health Canada accredits certain European notified bodies.

The next step will be to prepare and submit a Canadian Medical Device License (MDL) application for your device (Class II–IV devices). Compared to a US 510(k) application, MDL applications are more straightforward for Class II devices and about the same for Class III devices.[39] Class IV MDL applications are comparable to a US PMA application. For Class III and IV, the manufacturer needs to prepare and submit a premarket review document. As already mentioned, Class I medical device manufacturers apply for a Medical Device Establishment License (MDEL) instead of MDL.

Health Canada reviews MDL application (Class II, III, and IV) and premarket review document (Class III and IV only).

---

[39]EMERGO, Canada medical device approval chart. Retrieved from https://www.slideshare.net/emergogroup/emergo-group-canada-regulatory-chart

If approved (Class II–IV), the device can be legally marketed. Licenses do not expire. However, to maintain the license, annual fees have to be paid to Health Canada. Failure to pay fees will result in license revocation.

Canada has implemented the Medical Device Single Audit Program (MDSAP). It is a single audit performed by an accredited Auditing Organization to meet quality management system requirements for multiple regions (Canada, Australia, Brazil, Japan, USA), derived from ISO 13485:2003

Like other countries, Canada deploys a simplified and expedited procedure for premarket approvals in public health emergencies. In particular, during COVID-19, the Ministry of Health issued an "Interim order respecting the importation and sale of medical devices for use in relation to COVID-19."[40]

Key Takeaways:

- The Canadian MedDev certification system combines both US and EU features: (a) rule-based, but (b) centralized. Similar to the EU, it relies on a network of third-party organizations (registrars) for an audit of QMS. However, the application is reviewed by a central authority (Health Canada).
- Manufacturers need to have a QMS system according to ISO 13485 with some additional country-specific requirements.
- Every Class II, III, IV device must be licensed. For Class III and IV devices, premarket authorization is required.
- A manufacturer is subject to annuals audits by a registrar.
- The license does not expire. However, annual fees have to be paid to Health Canada to maintain the license.

2.4   Australia

Australia has a regulatory system, which is quite similar to Canada's. The classification system is rule-based, and the market is regulated by a centralized authority, the Therapeutic Goods Administration (TGA), a subdivision of the Department of Health.

---

[40]Health Canada (2020), Interim order respecting the importation and sale of medical devices for use in relation to COVID-19. Retrieved from https://www.canada.ca/en/health-canada/services/drugs-health-products/drug-products/announcements/interim-order-importation-sale-medical-devices-covid-19.html

Similar to the EU, the Australia MedDev classification consists of six buckets:

- Class I non-sterile, non-measuring function,
- Class I sterile,
- Class I measuring function,
- Class IIa,
- Class IIb, and
- Class III.

The classification procedure is rule-based and can be determined using Schedule 2 of the Australian Therapeutic Goods (Medical Devices) Regulations 2002. If the device has European CE Marking, the classification will likely be the same. A CE Marking certificate from a notified body is generally accepted by the Therapeutic Goods Administration (TGA) as part of the registration.[41]

The first step of the registration process is to establish a local presence by appointing an Australian sponsor. The sponsor's name must appear on the device and labeling.

The next step will be to prepare an Australian Declaration of Conformity. The sponsor will require a current technical file or design dossier for these purposes.

On the next step, the sponsor submits the Manufacturer's Evidence (e.g., CE Marking certificate) in TGA Business Services (TBS) system for TGA's review and acceptance.[42]

Then, the sponsor submits Medical Device Application and pays fees. The Application includes Intended Purpose statement, classification, and Global Medical Device Nomenclature (GMDN) code

The TGA reviews the application. For Class III medical devices, the TGA performs an additional Level 2 audit of the Design Dossier (a Level 2 application audit is required for all Class III devices and a small percentage of Class IIb devices).

If approved, an Australian Register of Therapeutic Goods (ARTG) listing number will be issued (ARTG Certificate of Inclusion). The listing will be included in the ARTG database on the TGA website.

---

[41]EMERGO, Australia TGA Approval Process for Medical Devices. Retrieved from https://www.emergobyul.com/resources/australia-process-chart
[42]EMERGO, Australia TGA Approval Process for Medical Devices. Retrieved from https://www.emergobyul.com/resources/australia-process-chart

Registrations do not expire as long as there are no changes to the device that would invalidate the ARTG listing, a current CE Marking certificate (if applicable) is on file with the TGA, and the annual ARTG listing fee is paid.[43]

In the presence of CE Marking, Class I–II devices' registration process is inexpensive and can be performed within several months. The application audit (required for Class III and certain Class IIb devices) may significantly increase timelines. So, 7–14 months are more realistic in this case.

Key Takeaways:

- The Australian MedDev certification system relies heavily on European certifications. The TGA generally accepts a CE Marking certificate from a notified body as part of the registration.
- Every medical device must be registered.
- Registration does not expire. However, annual ARTG listing fees have to be paid to maintain the registration.

2.5   Japan

Japan probably has the most complex regulatory system out of the five primary MedDev markets we discussed. However, it is the second largest in the world, ahead of China and Germany.

Japanese MedDev market is regulated by the Pharmaceuticals and Medical Devices Agency (PMDA). Classification of a medical device in Japan can be determined based on the Pharmaceuticals and Medical Devices Act (PMD Act) and the Japanese Medical Device Nomenclature (JMDN) codes.

Medical devices in Japan are split into five regulatory buckets based on their risk-based classification

- Class I – General medical devices,
- Class II – Specified controlled medical devices,
- Class II – Controlled medical devices,
- Class III – Highly controlled medical devices, and
- Class IV – Highly controlled medical devices.

---

[43]EMERGO, Australia TGA Approval Process for Medical Devices. Retrieved from https://www.emergobyul.com/resources/australia-process-chart

Each bucket has its own regulatory pathway. However, similarly to the US market, there are certain exceptions to that regulatory pathway. For example, if the device is new (there are no existing JMDN codes), certain higher-class requirements can be applied. Or some Class III medical devices are considered "Specified Highly Controlled" devices and follow the same approval route as Class II "Specified Controlled" devices. Another important aspect of the Japanese market is that all documentation needs to be in Japanese.

The first step in the regulatory approval process will be to establish a local presence by appointing a Marketing Authorization Holder (MAH).

The next step differs for domestic and foreign manufacturers. Japanese manufacturers must register domestic facilities with local prefectural authorities. Foreign manufacturers must submit a Foreign Manufacturer Registration (FMR) application to the Pharmaceuticals and Medical Devices Agency (PMDA).

All applicants must have a quality management system (QMS) that complies with the PMD Act and the Ministry of Health, Labour, and Welfare (MHLW) Ordinance #169. Ordinance #169 is based on ISO 13485; however, it has certain country-specific requirements.

All Class II and III device manufacturers need to have their QMS systems audited. For Class II–specified controlled medical devices, the audit will be performed by a Registered Certification Body (RCB), a third party registered with the Pharmaceuticals and Medical Devices Agency. All other device manufacturers need to be audited by PMDA.

QMS audit is part of the application review process. There are three types of premarket applications.

For Class I devices, premarket submissions must be submitted to PMDA.

For Class II–specified controlled medical devices, a premarket certification application must be submitted to a Registered Certification Body (RCB).

For all other types of devices, the Premarket Approval application and registration dossier must be submitted to PMDA.

During the regulatory review, the QMS system will get QMS certification (RCB or PMDA, respectively). Premarket certificate will be issued by the RCB or Premarket Approval certificate issued by the Ministry of Health, Labour, and Welfare.

QMS system needs to be audited every five years. Approvals do not expire, but Foreign Manufacturer Registration (FMR) renewal fees need to be paid to maintain their validity.

Key Takeaways:

- The Japanese regulatory system is considered challenging due to its complexity and Japanese language requirements. However, it is the second-largest MedDev market in the world.
- Manufacturers need to have a QMS system according to ISO 13485 with some additional country-specific requirements.
- Class II–IV medical devices require premarket certification (Class II–specified controlled medical devices) or Premarket Approval (all other types)
- Approvals do not expire. However, fees have to be paid to maintain the approvals.

2.6   Other Countries

The five markets, which we briefly discussed above, are often referred to as "Golden Billion." They are well-established markets with predictable market structures. However, with the globalization in recent years, other lucrative MedDev markets emerged: China, India, Brazil, Mexico, Asia Pacific, and the Middle East and Gulf countries.

These markets are in different stages of regulatory development. So, it is challenging to provide any guidance for them. We picked just two markets: China and the Asia Pacific. The first was chosen due to its tremendous market opportunities. The second (Asia Pacific) was selected due to its relatively well-established regulatory framework.

2.6.1   China

China is one of the largest and most dynamic MedDev markets in the world. So, it represents a vast market opportunity. However, it is one of the most complicated markets as well. The regulations are strict, immense, and change very rapidly. The fast pace of changes is the primary reason we do not consider Chinese market regulations in more detail. Probably, the information will be completely outdated by the time of publishing this book.

Nevertheless, it is helpful to understand the basics of this system. The Chinese regulatory system was modeled based on the US FDA. Probably, not coincidentally, initially, it was called CFDA.

The Chinese medical device classification is risk-based and consists of three classes: I, II, and III. However, there are significant differences in the application process comparing to the US market. Notably, there are substantial differences between foreign and local manufacturers.

For local manufacturers, there is a need to perform clinical performance validation in two clinical studies. The requirements for these clinical studies are quite different from other countries: (a) they need to be completed in China (foreign studies are not accepted), and (b) they need to be performed in certain (accredited) hospitals.

For foreign manufacturers, the submission threshold is much higher. Before submitting to CFDA, the device needs to be cleared/ approved in the home country. The manufacturer needs to demonstrate proof of home country approval using documentation such as a Certificate of Free Sale (CFS) or Certificate to Foreign Government (CFG).[44] After that, the device still needs to be tested by a CFDA authorized Medical Device Evaluation Center. For devices not on a clinical trial exemption list, China Clinical Evaluation must be performed through two local clinical studies in China. Thus, entering the Chinese market from outside for Class II and III devices is a multi-stage multi-year process.

2.6.2   Asia Pacific

The Asia Pacific region (Singapore, Thailand, Malaysia, Indonesia, and Vietnam) is another exciting market opportunity. These are dynamic countries with rapidly evolving healthcare systems (other than Singapore, which has a well-established healthcare system). These countries try to coordinate and harmonize their MedDev standards. However, they are still somewhat different. The common theme between these markets is that they accept medical device certifications from five major markets (United States, Canada, EU, Australia, and Japan). It means that if you have a MedDev clearance/

---

[44]EMERGO, China medical device approval chart. Retrieved from https://www. slideshare.net/emergogroup/emergo-group-china-regulatory-chart

approval in one of these primary markets, then the certification in South-East Asian countries can be significantly simplified. However, there is a certain nuance here as well. At least some countries (e.g., Thailand) require the Certificate of Free Sale, which means you need to be certified in your home country. It means that a German company with an FDA clearance, but not an EU clearance, will not benefit from this simplified procedure.

Chapter 3

Prerequisites

In order to get to the regulatory approval/clearance stage, the device manufacturer needs to complete several necessary preliminary steps. Firstly, a quality management system (QMS) needs to be implemented company-wide. Similarly, risk management needs to be adopted. Secondly, for the device, the manufacturer needs to cover biocompatibility, electromagnetic compatibility and electrical safety, and performance testing, among other things. Performance testing may require clinical testing in some cases. Thus, we can consider these elements as prerequisites.

It should be noted that most of these steps are the direct consequence of our decision to develop a medical device. Most of these steps are not required for any other hardware project. So, they can be considered as a self-inflicted burden. These prerequisites will be the scope of the current chapter.

3.1 Quality Management System

QMS requirements are applicable to companies involved in one or more stages of the medical device lifecycle, including design and development, production, storage and distribution, installation, servicing, and final decommissioning/disposal of the device. It is also applicable to companies involved in the design, development, and provisioning of associated activities (e.g., technical support).

Bringing a Medical Device to the Market: A Scientist's Perspective
Gennadi Saiko
Copyright © 2022 Jenny Stanford Publishing Pte. Ltd.
ISBN 978-981-4968-25-6 (Hardcover), 978-1-003-31221-5 (eBook)
www.jennystanford.com

In most countries, QMS requirements are based on ISO 13485 standard (although with some country-specific deviations). In the United States, the QMS requirements are imposed by the FDA Quality System Regulation (QSR) found in 21 CFR Part 820, commonly known as the FDA Good Manufacturing Practice (GMP).

There are several important concepts in the QMS regulations.

Firstly, quality management is process-based. What is the process? Any activity that receives input and converts it to output can be considered a process. Thus, the organization needs to identify (or design), document, and manage multiple interlinked processes impacting product quality and safety.

Obviously, not all organizational processes must be included in the QMS. Only the key processes essential for the product's quality and safety need to be covered by the QMS. However, this list is quite extensive. To have compliant QMS, the company should have procedures for:

- Product planning,
- Customer-related processes (e.g., requirements specified by customers, user training),
- Design and development,
- Purchasing (including criteria for the evaluation and selection of suppliers),
- Production and service provision, and
- Control of monitoring and measuring equipment.

Quality-related processes need to be managed properly. According to clause 4.1.3 of ISO 13485 standard for each QMS process, the organization "shall:

(a) Determine criteria and methods needed to ensure that both the operation and control of these processes are effective;

(b) Ensure the availability of resources and information necessary to support the operation and monitoring of these processes;

(c) Implement actions necessary to achieve planned results and maintain the effectiveness of these processes;

(d) Monitor, measure as appropriate, and analyze these processes;

(e) Establish and maintain records needed to demonstrate conformance ... and compliance..."

Secondly, quality management is expected to be risk-based. It means that the risks have to be assessed and managed. We will discuss it later in this chapter.

The key approach to quality management is proper documentation. In particular, according to clause 4.1.1 of ISO 13485, "The organization shall establish, implement and maintain any requirement, procedure, activity or arrangement required to be documented by this International Standard."

What happens if the company is involved in several steps of the product lifecycle and outsource everything else? When the company outsources any process that affects conformity to requirements, the company retains the responsibility of conformity. Thus, the company needs to monitor and ensure control over such processes. It can be achieved through quality agreements (in writing).

While most of the mentioned above key processes are performed by a limited number of people (e.g., procurement), design and development is typically at the heart of the company's operations and starts early in the startup's life. Thus, we will consider applications of QMS requirements for design and development in more detail.

3.1.1   Design Control

Control over product design is a requirement for the QMS and is intended to follow the product's life (including software) across the entire product lifecycle. It is achieved through the implementation of design controls.

Design controls are a set of quality practices and procedures incorporated into the design and development process. They control the design process to assure that device specifications meet:

- User needs and
- Intended use

Regulatory agencies have a list of requirements linked to design and development. In particular, the design and development process is split into several phases:

- Design and development planning,
- Design input,
- Design output,

- Design review,
- Design verification, and
- Design validation.

Design Inputs

User requirements, which are used for the product design, are an example of design inputs. However, the design input concept is broader. It includes things beyond user requirements. Among other things, Design Inputs comprise:

- Functional, performance, usability, and safety requirements,
- Applicable regulatory requirements and standards, and
- Risk management outputs.

Design Output

Design output is the design specifications that should meet design input requirements, as confirmed during design verification and validation, and ensured during design review.

The output includes the device, its labeling and packaging, associated specifications and drawings, and production and quality assurance specifications and procedures.

Design outputs must contain and/or refer to "acceptance criteria" essential for the proper functioning of the device.

Design Review

To manage and control design outputs, the company needs to perform Design Reviews. Design reviews are systematic reviews, which aim to:

- Evaluate the ability of design outcomes to meet requirements and
- Identify and propose necessary actions to address identified gaps.

Design Verification

The design verification step aims to answer the question, "Did we make the product right? (i.e., specified requirements have been fulfilled)." It confirms that design outputs meet design input requirements. Types of verification activities include:

- Inspections,
- Tests, and
- Analysis.

Design Validation

The design validation step aims to answer the question, "Did we make the right product? (i.e., device specifications conform with the user needs and intended use(s))." Design validation needs to be performed to conform to defined user needs and intended uses:

- Under actual or simulated use conditions,
- Under defined operating conditions, and
- On initial production units, lots, or batched or their equivalents.

Design Validation includes software validation and risk analysis, where appropriate.

In Fig. 3.1, one can see how all elements of the design and development process are integrated.

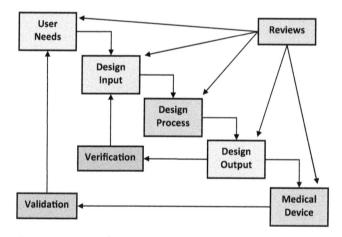

Figure 3.1 Integration of elements of the design and development process.

It should be mentioned that ISO 13485 and QSR do not prescribe or endorse a particular project management methodology. The company can choose any (e.g., Waterfall or Agile) and build in required design controls in it.

3.1.2 Core Components of QMS

In addition to being broad in scope, QMS requirements are pretty deep. It is insufficient to write procedures. You have to follow these procedures and create a paper trail (records) to prove that you comply with them.

What are the core components of the QMS? According to the standard, the company should have:

- Quality manual
- Medical device file
- Control of documents
- Control of records

In a nutshell, to have a compliant QMS, the company needs to have a quality manual and policies and procedures in place. It should also have version control for policies, procedures, and other important documents. Finally, all steps need to be documented and records retained. This high-level approach is depicted in Fig. 3.2.

Figure 3.2   Elements of quality management system.

3.1.3 Software Development

ISO 13485 considers software one of four generic product categories: services (e.g., transport), software, hardware, and processed materials (e.g., lubricant). Thus, the software is subject to the same standards as traditional devices.

Similarly, in the US, for medical devices in which software is a component of the device function, subpart C of the QSR requires that design controls be established for the device and its associated software.

An international standard has been developed to address specific software development questions: IEC 62304 medical device software—software life cycle processes.

In software design, controls start with identifying user requirements, followed by developing the product to meet those requirements, coding, integration, verification testing, and deployment.

The QSR in general and design controls, in particular, do not define a development process or prescribe a specific development methodology or a particular software development lifecycle; instead, they specify controls that must be integrated into the manufacturer's development processes.

Key Takeaways:

- QMS requirements are applicable to companies of all sizes.
- QMS requirements are pretty broad and deep, requiring some time and effort to implement. Thus, you should allocate time for their implementation.
- QMS requirements vary between countries. If you plan to implement a QMS, it will be prudent to design and implement the QMS, which complies with regulations in all your key markets.
- The QMS needs to be tailored to the organization. It will be impractical to copy and paste somebody else system. It will be beneficial to hire an experienced QMS consultant who will develop the system tailored to your organization.

3.2   Risk Management

A risk-based approach is a pillar of any MedDev regulations. Similarly, risk management is an essential part of any QMS.

Risk management should be engraved into the QMS. The risk management process should be defined, documented, and implemented. And it should be followed during all phases of the product development lifecycle.

The requirements for risk management are outlined in the international standard ISO 14971. ISO 14971 provides a systematic explanation of relevant terms and definitions and defines a risk management process. However, the standard does not set specific requirements. Instead, the standard explains the requirements, expectations, and stages of a risk management process. More detailed recommendations and guidance on the application of ISO 14791 can be found in ISO/TR 24971—Guidance on the application of ISO 14971. More specific requirements for risk management in software development can be found in the IEC 62304 standard.

3.2.1 Risk Management Process

The core concept of risk management is a hazard. The hazard is defined as a potential source of harm. Each hazard can be characterized by severity and the probability of its occurrence. This combination of the likelihood of occurrence of harm and the severity of that harm is called a risk. Risks need to be analyzed and evaluated.

Risk management's primary approach is to identify hazardous situations, evaluate their risks, and design/implement risk controls to manage these risks.

The risk management process consists of several steps. It includes:

- Risk management planning, including establishing acceptance criteria
- Risk analysis
- Evaluating risk for each hazardous situation
- Developing risk controls whenever the risk must be reduced
- Evaluating the residual risk
- Risk management review
- Production and post-production information

This process can be iterative. If the residual risk is higher than the acceptance criteria, then additional risk controls need to be designed and implemented.

Schematically, the risk management process is depicted in Fig. 3.3.

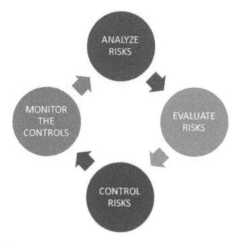

Figure 3.3   Risk management process.

Severity

To categorize the severity of the hazard posed by a medical device, the adverse effects on human health are used. A common technique that is used is to define descriptions for various levels of severity. The severity levels and descriptions should be aligned with the medical device. For example, according to IEC 62304 (which is used for the classification of risks posed by software), the following severity classification can be used:

- Minor (Class A): No injury or damage to health is possible
- Medium (Class B): **Non-serious injury** is possible
- High (Class C): Death or **serious injury** is possible

Here, "**serious injury**" means injury or illness that directly or indirectly: (a) is life-threatening, (b) results in permanent impairment of a body function or permanent damage to a body structure, or (c) necessitates medical or surgical intervention to prevent permanent impairment of a body function or permanent damage to a body structure.

"Permanent impairment" in this classification means an irreversible impairment or damage to a body structure or function, excluding trivial impairment or damage.

Severity categories can be more granular and contain, for example, five categories: critical, major, serious, minor, negligible.

Likelihood/Probability of Hazard

The hazard is also characterized by the probability (or likelihood) of its occurrence. Typically, it is difficult to quantify this probability precisely. In practice, the categorical approach is used instead. The standard approach is to use at least three bins (e.g., high, medium, and low probabilities). Occasionally, a more granular (four or five bins) hazard likelihood classification has been used. For example, the following classification can be used:

- Frequent—1 in 100
- Probable—1 in 1,000
- Occasional—1 in 10,000
- Remote—1 in 100,000
- Improbable—1 in 1,000,000

Typically, the likelihood of occurrence might include quantitative terms (as in the example above). For new products, it can be difficult to estimate because little data are available. In this case, a little bit less quantitative categorization can be used. An example of such categorization can be:

- Frequent—almost on every use of the device
- Probable—occurs the majority of times, but not always
- Occasional—fall in between probable and remote occurrences
- Remote—more than one occurrence a year, but still unlikely
- Improbable—isolated occurrences. Less than one event a year

Note that there is no definitive guidance on severities and the likelihood of harm. The severity and likelihood levels and descriptions should be aligned with the medical device. When estimating severity and occurrence for harms of each hazardous situations you should leverage objective evidence to support your estimates.[1] They should be documented in the Risk Management Plan.

---

[1]Speer J (2020). The Definitive Guide to ISO 14971 Risk Management for Medical Devices. Retrieved from https://www.greenlight.guru/blog/iso-14971-risk-management

Risk Evaluation

If the severity and probability are defined, then the risk levels can be assessed using a risk matrix or risk acceptability matrix, a standard tool in risk management. An example of a risk matrix for a medical device is represented in Table 3.1.

Table 3.1   An example of a risk matrix for a medical device

		Severity of harm				
		Negligible	**Minor**	**Serious**	**Critical**	**Catastrophic**
Probability of Occurrence	Frequent	Medium	High	High	High	High
	Probable	Medium	Medium	High	High	High
	Occasional	Low	Medium	Medium	High	High
	Remote	Low	Low	Medium	High	High
	Improbable	Low	Low	Low	Medium	Medium

The advantage of the risk matrix is that it allows visualization. Each cell of the risk matrix is assigned (and color-coded for visualization) to a risk level. In the case above, there are three risk levels: low, medium, and high.

A common practice for risk evaluation is to identify which risk levels are acceptable and which require risk reduction. It needs to be defined in the Risk Management Plan.

The US market's typical practice is to correlate the low-risk levels with acceptable risk and the high-risk levels with unacceptable risk. The medium-risk level often fits into what is referred to as "as low as reasonably practicable." Items in the high-risk levels require risk reduction, and those in the medium level are generally considered for risk reduction, as well.

For the EU market, the approach needs to be different, such as that the EU MDR states that you must "reduce risks as far as possible," meaning you need to consider risk reductions for all risks, regardless of the risk level.

Risk Controls

Risk controls are measures taken to reduce the risk to an acceptable level. While multiple risk control options can be typically available, their application should be considered according to the following priority:[2]

- Inherent safety by design,
- Protective measures in the actual medical device and/or manufacturing process,
- Information for safety, such as labeling and instructions for use.

Thus, the labeling and instructions for use should not be overused and treated as a last resort.

Once risk controls are designed, they need to be implemented. Once implemented, they need to be verified, and their effectiveness needs to be determined. Records of this shall be documented.

Residual Risks

Once risk controls are implemented and verified for effectiveness, a residual risk still may exist. Residual risk needs to be evaluated. If it is still unacceptable (or if following EU MDR is not reduced as far as possible), it needs to be addressed.

What happens if you tried to reduce risks using risk controls but some of them still are at an unacceptable level? The solution here can be to conduct and document a benefit-risk analysis (BRA), which is a special provision for moving forward with unacceptable risks.

The BRA must be documented and provided objective evidence and rationale for why the medical benefits outweigh the unacceptable risks.[3] However, you should remember that financial reasoning should never be included in the BRA.

Once all individual risks have been evaluated, the next step is to assess whether the medical device's overall residual risk in its whole entirety is acceptable. It needs to be done using the same severity, occurrence, risk level, and risk acceptability criteria used throughout the individual risk evaluation process.

---

[2]Speer J (2020). The Definitive Guide to ISO 14971 Risk Management for Medical Devices. Retrieved from https://www.greenlight.guru/blog/iso-14971-risk-management

[3]Speer J (2020). The Definitive Guide to ISO 14971 Risk Management for Medical Devices. Retrieved from https://www.greenlight.guru/blog/iso-14971-risk-management

Risk Management Review

Before going to market, the results of all steps of the risk management process shall be reviewed to ensure completeness. It is done through a Risk Management Report, which summarizes all risk management activities and includes any benefit-risk analyses and explanation of overall risk acceptability.

Production and Post-Production Information

The risk management process does not stop with a product launch. It needs to be updated on information from use, including claims and customer feedback.

3.2.2   Risk Management Procedures

Above, we discussed the essential components of the risk management process. What needs to be implemented to become compliant? To be compliant with the risk management requirements, each device must have its own Risk Management Plan And Risk Management File.

Risk Management Plan

A Risk Management Plan is a product-level document. The Risk Management Plan should identify the risk management activities anticipated and planned throughout the product's life cycle.

The Risk Management Plan is used to:[4]

- Define the scope of risk management activities.
- Assign risk management responsibilities and authorities.
- Specify management review requirements.
- Establish risk acceptability criteria for each plan.
- Describe how verification activities will be done for each medical device.
- Explain how production and post-production information for each medical device will be collected and reviewed.

---

[4]Praxiom, ISO 14971 Medical Device Risk Management in Plain English. Retrieved from https://www.praxiom.com/iso-14971.htm

Risk Management File

The primary use of a Risk Management File (RMF) is to establish traceability. RMF is the place to keep all risk management activities, documentation, and records. The RMF should contain evidence of:

- Risk management plan,
- Risk analysis,
- Risk evaluation,
- Risk controls,
- Evaluation of overall risk acceptability,
- Risk management review, and
- Production and post-production risks.

For example, the RMF needs to contain:

- records of risk analysis results for each identified hazard,
- records of risk evaluation results for each identified hazard,
- records of risk control measures and results for each identified hazard, and
- records of residual risk evaluation results for each identified hazard.

A best practice is to keep the product RMF content together in a single location for ease of access and use.

3.2.3   Risk Management and Design Controls

As you may notice, risk management (particularly risk controls is very similar in nature to other components of the QMS, particularly design controls. Is it duplication? Can design controls replace risk management?

The answer is "no." Good design controls do not replace risk management. These are complementary systems, which work together to improve patient safety and product quality.

In fact, they need to be intervened. For example, outputs of the risk management (hazards and hazardous situations) serve as design inputs.

3.2.4 Risk Management Roles and Responsibilities

In addition to following a systematic approach to risk management, the risk management roles should be defined and responsibilities assigned. Personnel, which perform risk management tasks, need to have appropriate knowledge and experience.

Medical device risk management also requires top management involvement. It requires that a company establishes a risk management policy. In addition to that, the company's management is responsible for:

- making sure there are adequate and appropriate resources for conducting risk management activities,
- ensuring the company's risk management processes are adequate and effective,
- reviewing the company's risk management processes for effectiveness.

For example, a good practice will be if the company's executive management approves the Risk Management Report.

3.3 Biocompatibility

Biocompatibility addresses the patient's safety. It is known that some materials may cause adverse health effects, e.g., adverse tissue reaction, necrosis, chronic inflammation, and blood clotting. They also may cause shock and cancer.

Biocompatibility becomes a more prominent issue nowadays. It is a critical section in any regulatory submission. Also, it a common topic in deficiency letters from regulators or notified bodies. Thus, it is worth considering some biocompatibility basics here.

While biocompatibility may be perceived as an issue related to the material only, in fact, it is not. Regulators do not approve/clear a specific material. They approve/clear the device, which contains a particular material. Thus, the regulators have a holistic view of biocompatibility. The biocompatibility covers the materials used and their processing methods, shapes, surfaces, etc.

US and EU regulators use quite similar approaches to address biocompatibility. The biocompatibility risk assessment relies on an ISO 10993-1 standard.[5] In particular, the FDA has a guidance document on the application of this standard.[6]

Despite strong similarities, there are certain differences. For example, EU requirements are stricter. They require additional information, including chemical information on raw materials and processing materials. In particular, according to MDR,[7] "Devices... shall only contain the following substances in a concentration that is above 0.1 % weight by weight (w/w) where justified pursuant to Section 10.4.2...substances which are carcinogenic, mutagenic or toxic to reproduction ('CMR')... substances having endocrine-disrupting properties for which there is scientific evidence of probable serious effects to human health." To prove that the concentrations of materials used in downstream processes are below 0.1% w/w, extensive documentation from the material manufacturer may be required.

The adverse reaction mechanisms can be split into certain biological aspects or biocompatibility endpoints. There are 14 of them in total, with 11 used often. Most commonly cited are cytotoxicity, genotoxicity, and carcinogenicity. In each particular case, there is no need to address all 14 aspects. Typically, it is required to discuss only those, which are relevant to a specific use case.

There are specific tools for assessing biocompatibility:

- Biological testing (animals and *in vitro*),
- Chemical testing,
- Risk assessment, and
- Rationales (e.g., why other tools mentioned above are not needed).

In most cases, the regulatory submission may contain a combination of several such tools (e.g., chemical testing in combination with a risk assessment).

---

[5]ISO 10993-1 Biological evaluation of medical devices — Part 1: Evaluation and testing within a risk management process. Retrieved from https://www.iso.org/standard/68936.html

[6]FDA, Use of International Standard ISO 10993-1. Retrieved from https://www.fda.gov/media/85865/download

[7]MDR Annex 1: GSPR #10, 10.4.1

Important Takeaways:

- Biocompatibility testing may cost >$100k and take months or years. Thus, start considering it early in the project
- Regulatory submission should include both materials and downstream processes discussion
- The general advice is to use known materials wherever it is possible. These materials have known properties, documentation and will attract less regulatory scrutiny.

3.4 EMC and Safety Testing

Medical devices need to be safe and efficient. That is why there are multiple standards, which address their safety.

Electromagnetic compatibility testing (EMC) and electrical safety are essential safety tests for electrical medical devices and are conducted in certified testing facilities.

The primary standard addressing medical device safety is IEC 60601-1—3rd edition—Medical electrical equipment—Part 1: General requirements for basic safety and essential performance.

In addition to the primary standards, certain aspects of safety and performance are defined by the collateral standards (numbered 60601-1-X).

For example, electromagnetic compatibility is addressed in the IEC 60601-1-2—4th edition—Medical electrical equipment— Part 1-2: General requirements for basic safety and essential performance—Collateral Standard: Electromagnetic disturbances.

A medical device needs to be designed with human factors in mind, which we will discuss later in this chapter. The medical device's usability needs to be tested according to IEC 60601-1-6 (Medical electrical equipment—Part 1–6: General requirements for basic safety and essential performance—Collateral Standard: Usability).

Most devices are for use in medical settings. However, some devices can be used in home settings by a patient. If this is the case, then the device may need to meet some additional requirements and get tested according to IEC 60601-1-11, Medical electrical equipment—Part 1–11: General requirements for basic safety and essential performance—Collateral standard: Requirements for medical electrical equipment and medical electrical systems used in the home healthcare environment.

If the medical device contains an alarm, then it needs to meet requirements and get tested according to IEC 60601-1-8, medical electrical equipment—Part 1–8: General requirements for basic safety and essential performance—Collateral Standard: General requirements, tests, and guidance for alarm systems in medical electrical equipment and medical electrical systems.

In addition to IEC 60601-1-X collateral standards, there are multiple other product-specific standards, which need to be tested against.

For example, suppose the device contains laser sources. In that case, it needs to meet requirements and be tested according to IEC 60601-2-22, Medical electrical equipment—Part 2-22: Particular requirements for basic safety and essential performance of surgical, cosmetic, therapeutic, and diagnostic laser equipment.

Suppose the device contains non-laser light sources (e.g., LEDs). In that case, it needs to be tested according to IEC 60601-2-57, Medical electrical equipment—Part 2-57: Particular requirements for the basic safety and essential performance of non-laser light source equipment intended for therapeutic, diagnostic, monitoring, and cosmetic/aesthetic use.

Photobiological safety of light sources is covered by IEC 62471, Photobiological safety of lamps and lamp systems.

EN 45502 covers surgery—Active implantable medical devices. Short-wave therapy is covered by IEC 60601-2-3 Edition 3.0 "Medical electrical equipment—Part 2–3: Particular requirements for the basic safety and essential performance of short-wave therapy equipment."

Nerve and muscle stimulators are covered by IEC 60601-2-10 Edition 2.0 "Medical electrical equipment—Part 2–10: Particular requirements for the basic safety and essential performance of nerve and muscle stimulators."

Endoscopic equipment is addressed in IEC 60601-2-18 Edition 3.0 "Medical electrical equipment—Part 2–18: Particular requirements for the basic safety and essential performance of endoscopic equipment."

In addition to that, the accredited testing facility may conduct a risk assessment according to ISO 14971.

Below, we provide some basic information on the two most common tests performed by accredited facilities: electromagnetic compatibility testing and electrical safety testing.

3.4.1 Electromagnetic Compatibility Testing

Electromagnetic compatibility (EMC) is the electrical equipment and systems' ability to function acceptably in their electromagnetic environment.

All electrical medical devices need to be tested for EMC. There are multiple standards, which can be applicable to EMC testing. The most relevant standards for medical devices are:

- IEC 60601-1-2 Edition 3: 2007: Medical Electrical Equipment— Part 1–2: General Requirements for Safety—Collateral Standard: Electromagnetic compatibility – Requirements and Tests
- AAMI/ANSI/IEC 60601-1-2: 2007/(R)2012: Medical Electrical Equipment—Part 1–2: General Requirements for Safety—Collateral Standard: Electromagnetic Compatibility – Requirements and Tests
- IEC 60601-1-2 Edition 4.0:2014: Medical Electrical Equipment, Part 1-2: General Requirements for Basic Safety and Essential Performance—Collateral Standard: Electromagnetic Disturbances – Requirements and Tests
- AAMI/ANSI/IEC 60601-1-2: 2014: Medical Electrical Equipment— Part 1–2: General Requirements for Safety— Collateral Standard: Electromagnetic disturbances – Requirements and Tests.

Before Jan 1, 2019, medical devices in most jurisdictions needed to comply with the 3rd Edition of IEC 60601-1-2. After that date, typically, medical devices need to comply with the 4th Edition. The notable exception is the United States, where legacy devices still can comply with the 3rd Edition.

Conceptually, the EMC tests aim to make sure that (a) the device does not cause other equipment to fail in the presence of its unwanted emissions, which are known as radio frequency interference (RFI), (b) the device can function correctly in the presence of RFI (immunity).

Based on that, EMC tests divide broadly into emissions testing and susceptibility/immunity testing.

EMC testing is conducted in a specialized EMC test chamber.

Emissions are typically measured for radiated field strength and, where appropriate, for conducted emissions along cables and wiring.

Radiated field susceptibility testing typically involves a high-powered RF energy source and a radiating antenna to direct the energy at the device under test (DUT).[8]

Susceptibility testing also includes electrostatic discharge (ESD) testing. Electrostatic discharge testing is typically performed with a piezo spark generator called an "ESD pistol."[9]

3.4.2   Electrical Safety Testing

Electrical safety testing is carried out to evaluate the potential risks of electrical shocks to customers when using their products. This testing is conducted by an accredited laboratory and can be performed in a specialized facility or at the manufacturer's site.

For medical devices, the most relevant standard is:

- IEC 60601-1—3rd edition—Medical electrical equipment—Part 1: General requirements for basic safety and essential performance.

Electrical safety testing usually includes the following tests:

- A high voltage test (dielectric withstand test) measures an electrical product's ability to withstand a high voltage applied between a product's electrical circuit and the ground.
- The leakage current test evaluates whether the current that flows between an AC source and the ground does not exceed a specified limit.
- The insulation resistance test calibrates the quality of the electrical insulation used.
- Ground continuity test ensures that a clear path is available between all exposed metal surfaces and the power system ground.

---

[8]EMC Testing Labs - Electrical Safety Testing Laboratory. Retrieved from https://www.itcindia.org/emc-testing/

[9]EMC Testing Labs - Electrical Safety Testing Laboratory. Retrieved from https://www.itcindia.org/emc-testing/

3.4.3 Special Cases

There are several standard technologies, which used in many medical devices. Below, we will briefly discuss the two most common ones (wireless technology and rechargeable batteries) and their implications on medical device safety.

3.4.3.1 Wireless technology

The incorporation of wireless technology in medical devices can have multiple benefits. It includes increased patient mobility (no wires that tether a patient to a medical bed), the ability to remotely access and monitor patient data regardless of the location of the patient or physician (hospital, home, office, etc.), and provides the ability to remotely program devices or update the firmware. These benefits can significantly impact patient outcomes. However, they may pose significant challenges, including lost connection, data security, etc. Due to these known challenges (notably, quality of service, security, and electromagnetic compatibility (EMC)), wireless technology received special attention from the FDA.

The primary regulator of wireless technologies in the United States is the Federal Communications Commission (FCC), which oversees the use of the public RF spectrum. Thus, your company needs to be registered with the FCC if your product has a wireless function. Similarly, in Canada, the company needs to be registered with Industry Canada (IC).

FCC and IC registration is well described in footnote[10] "FCC registration is a two-step process. First, the company applies for its FCC Registration Number (FRN), a unique number associated with the company for all transactions. Registration is free and can be done online using the CORES or by completing FCC Form 160.

After receiving its FRN, the company must then apply for a Grantee Code, a unique three-character identifier which the company will use for all its wireless products. The FRN and Grantee Code are specifically associated with the company's address and with the individual designated as authorized signatory for the company."

---

[10]Laird, A Guide to the FCC and Industry Canada Certification Process, Retrieved from https://www.lairdconnect.com/resources/white-papers/guide-on-fcc-and-industry-canada-certification

To get the product to market, the company needs to obtain the FCC ID number for each specific product. To register the product with the FCC and obtain this number, the company needs to submit product-related information to the FCC. For example, documentation required for the filing includes the theory of operation, block diagram, schematics, and antenna datasheets, and user manuals. The primary part of the submission is test results from a certified lab. Note that if a pre-certified module is used in the device, only FCC Part B testing may be required for the FCC filing. Such as it can be integrated with testing for FDA, the whole process becomes faster and cheaper.

The FDA's policies on wireless medical devices are coordinated with the FCC. To address applications of wireless technologies in the healthcare environment, the FDA has issued a guidance document[11]

There are multiple standards, which can be applicable to medical devices with a wireless function:

- IEC 60601-1-2 Edition 3: 2007: Medical Electrical Equipment— Part 1-2: General Requirements for Safety—Collateral Standard: Electromagnetic Compatibility – Requirements and Tests
- IEC 60601-1-2 Edition 4.0: 2014: Medical Electrical Equipment, Part 1-2: General Requirements for Basic Safety and Essential Performance—Collateral Standard: Electromagnetic Disturbances – Requirements and Tests
- AAMI/ANSI/IEC 60601-1-2: 2007/(R)2012: Medical Electrical Equipment—Part 1-2: General Requirements for Safety—Collateral Standard: Electromagnetic Compatibility – Requirements and Tests
- AAMI/ANSI/IEC 60601-1-2: 2014: Medical Electrical Equipment—Part 1-2: General Requirements for Safety— Collateral Standard: Electromagnetic Disturbances – Requirements and Tests
- AAMI TIR18: Association for the Advancement of Medical Instrumentation—Guidance on electromagnetic compatibility of medical devices in healthcare facilities
- ANSI/IEEE C63.18-2014: American National Standard Recommended Practice for an On-Site, Ad Hoc Test Method

---

[11]FDA (2013). Radio Frequency Wireless Technology in Medical Devices, A guidance document. Retrieved from https://www.fda.gov/media/71975/download

for Estimating Electromagnetic Immunity of Medical Devices to Radiated Radio Frequency (RF) Emissions from RF Transmitters

- AIM Standard 7351731: Medical Electrical Equipment and System Electromagnetic Immunity Test for Exposure to Radio Frequency Identification Readers—An Aim Standard.

Key Takeaways:

- Manufacturers of medical devices with wireless function needs to be registered with the RF spectrum regulators (for example, the FCC in the United States)
- The easiest (and cheapest) way to comply with an extensive set of wireless technology standards is to use pre-certified modules and components.

3.4.3.2 Portable batteries

Batteries (and particularly, lithium-ion batteries) are another area of regulatory scrutiny. While being compact and having excellent energy storing capacities, they may pose both chemical and electrical hazards. Dangers include chemical burn, fire, and electrical shock.

Several standards and testing methods cover batteries. For batteries, the key standards are IEC 62133 (rechargeable batteries), the IEC 60086 suite (non-rechargeable batteries), IEC 61960, and IEC 62281 (the IEC version of UN 38.3). UL 1642 applies to lithium-based cells in the US, while battery packs are covered by UL 2054, which also references UL 1642 for lithium cells.

IEC 62133 "Secondary cells and batteries containing alkaline or other non-acid electrolytes—safety requirements for portable sealed secondary cells, and batteries made from them, for use in portable applications" is the primary standard for rechargeable batteries. IEC 62133 replaced UL 1642 standard in Europe and several other markets. But UL 1642 is still in use in the US.

The certification applies to cell and battery packs. For example, cells must be certified to IEC 62133 to certify the battery. However, battery packs—including those intended for use in medical products—must be additionally evaluated for full compliance to IEC 62133.

In addition to general safety standards, lithium batteries pose some logistical challenges. Lithium batteries are classified as dangerous goods (Class 9 material—termed "miscellaneous dangerous goods"). The transport of lithium batteries is subject to international regulations is covered by the United Nations (UN) regulations. The specific UN regulations covering the shipment of these batteries are as follows:

- UN 3090, Lithium metal batteries (shipped by themselves),
- UN 3480, Lithium-ion batteries (shipped by themselves),
- UN 3091, Lithium metal batteries contained in equipment or packed with equipment, and
- UN 3481, Lithium-ion batteries contained in equipment or packed with equipment.

To comply with multiple standards, batteries need to go through rigorous testing in a certified laboratory. For example, UN 38.3 presents a combination of significant environmental, mechanical, and electrical stresses designed to assess lithium batteries' ability to withstand the anticipated rigors incurred during transport.

Key Takeaway:

- The easiest way to comply with battery safety standards is to use medical-grade batteries with existing certifications (e.g., IEC 62133 and UL 1642). It will save a significant amount of time and money.

3.4.4 Standards and Accredited Labs

510(k) or PMA submissions to FDA require test reports provided by recognized labs. Similarly, test reports are required for inclusion in CE Technical File for Europe and the United States and Canadian certification mark for medical devices. These tests are performed in certified testing faculties.

When we talk about certification marks, the first, which comes to mind, is UL, at least here, in North America. UL LLC (formerly known as Underwriters Laboratories) is a global safety certification company. However, UL is not the only safety certification option available. Several companies were approved to perform safety testing by the US federal agency Occupational Safety and Health Administration (OSHA). Among other well-known names are Bureau

Veritas, Canadian Standards Association (CSA), ETL/Intertek, SGS, and TUV.

OSHA maintains a list of approved testing laboratories, which are known as Nationally Recognized Testing Laboratories.

Similarly, in Europe, there are multiple notified bodies, which typically serve as accredited testing facilities as well.

Accredited testing facilities conduct tests according to recognized standards. Where are these standards coming from? Standards can be split into national and international standards. In many cases, national standards are adaptations or localizations of international standards. In particular, the country may adopt an international standard with some additional requirements (National Differences) to address local safety concerns.

In general, international standards are developed by international standards organizations. The International Organization for Standardization (ISO), the International Electrotechnical Commission (IEC), and the International Telecommunication Union (ITU) are the three largest and most well-established standard organizations, which form the World Standards Cooperation (WSC) alliance. They have established numerous standards covering almost every possible topic. Many of these are then adopted worldwide, replacing various incompatible local standards. Some of these standards are naturally evolved from industry-based or country-based standards. In contrast, others have been developed from scratch by groups of experts from various technical committees (TCs).

ISO is composed of the national standards bodies (NSBs), one per member economy. The IEC is similarly composed of national committees, one per member economy.

In the United States, the standard development is coordinated by two organizations. The National Institute of Standards and Technology (NIST) is a non-regulatory agency of the United States Department of Commerce. The American National Standards Institute (ANSI) is a non-profit organization. They oversee the development of voluntary consensus standards for products, services, processes, systems, and personnel in the United States. They also coordinate US standards with international standards so that American products can be used worldwide.

ANSI accredits standards developed by representatives of other standards organizations, government agencies, consumer groups, companies, and others. These standards ensure that products' characteristics and performance are consistent, that people use the same definitions and terms, and that products are tested the same way. ANSI also accredits organizations that carry out product or personnel certification in accordance with requirements defined in international standards.[12]

3.4.5 CB Schema

With multiple certification agencies on the market, it seems tricky to get safety certifications in various countries. However, there is a solution to this problem. The CB Schema, established by the International Electrotechnical Committee for Conformity Testing to Standards for Electrical Equipment (IECEE), provides a means for the mutual acceptance of test reports among participating safety certification organizations in certain product categories.[13]

Each participating country (there are more than 50 of them, including primary MedDev markets) has one or more organizations accepted by the IECEE as National Certification Bodies (NCBs).

Each NCB can test the product to a set of standards (e.g., to IEC 60601 for Medical Electrical Equipment), with or without National Differences (National Differences are special requirements that the IECEE CB Scheme permits each country to adopt to address local safety concerns).

Thus, testing at one NCB may solve the certification problem for multiple markets. For example, NCB will test the product to all harmonized standards and any National Differences required for the countries you want to market your product. Then, the NCB will then issue a CB Test Certificate and a CB Test Report that you can use to obtain national certifications in participating countries.[14] The final step is to apply to an NCB in your target countries for national certifications.

---

[12]AWS, American National Standards Institute (ANSI), Retrieved from https://awo.aws.org/glossary/american-national-standards-institute-ansi/

[13]UL Canada, CB Scheme. Retrieved from https://canada.ul.com/ulcprograms/cbscheme/

[14]UL, UL Schemes and Certification Bodies. Retrieved from https://www.ul.com/resources/ul-schemes-and-certification-bodies

CB schema can be particularly useful if you consider Japan, South Korea, Russia, or Israel as your secondary market.

Key Takeaways:

- Medical devices are subject to many safety standards. Review standard landscape early in the design process to avoid gaps and costly mistakes.
- There are multiple safety certification options available. It can be beneficial to research them for the price, geographical coverage, and test scope.

3.5 Performance Testing

Certain performance testing must be performed to ensure a medical device's safety and effectiveness and results demonstrated to a regulator. The scope of testing varies significantly with the class of the medical device.

As you may notice before, some standards include language about essential performance. Essential performance refers to the performance necessary to achieve freedom from unacceptable risk. Essential performance is part of safety testing and is typically performed in accredited facilities. However, in addition to safety, regulators expect results, which support the efficiency of the device. It is generally referred to as performance testing.

Performance testing can be roughly split into non-clinical laboratory studies and clinical investigations. Not all premarket submissions require clinical studies. For example, less than 10% of 510(k) submissions require clinical data. However, PMA and De Novo submissions will almost certainly require clinical data.

3.5.1 Non-clinical Laboratory Studies

Non-clinical laboratory studies typically refer to non-clinical bench performance testing, human factors (which we consider later separately), animal studies, and studies evaluating *in vitro* diagnostic devices' performance characteristics.

Non-clinical bench performance testing may include mechanical and biological engineering performance (such as fatigue, wear, tensile strength, compression, and burst pressure); bench tests using *ex vivo*, *in vitro*, and *in situ* animal or human tissue; and animal carcass or human cadaveric testing.

To provide guidance on the bench testing content, which needs to be covered in the premarket submission, the FDA has issued a guidance document.[15] It provides FDA's recommendations on the content for non-clinical bench performance testing results, which need to be included in all types of premarket submissions [i.e., Premarket Approval (PMA), Humanitarian Device Exemption (HDE), Premarket Notification [510(k)], investigational device exemption (IDE) applications, and De Novo requests]. In particular, the premarket submission should include Test Report Summaries. The FDA guidance consists of a detailed list of all necessary sections of the Test Report Summaries. However, in addition to that, the FDA recommends that premarket submissions also include Complete Test Reports and Test Protocols. Even though there are no such sections in the submission, they can be attached to the submission's main body (e.g., in an appendix).

The premarket submission should also discuss how the non-clinical bench performance test results support the overall submission (e.g., substantial equivalence for 510(k) or reasonable assurance of safety and effectiveness for PMA or De Novo request).

The non-clinical laboratory studies section of a PMA submission typically includes information on microbiology, toxicology, immunology, biocompatibility, stress, wear, shelf life, and other laboratory or animal tests.

PMA submissions require extensive safety testing. Non-clinical studies for safety evaluation must be conducted in compliance with 21 CFR Part 58 (Good Laboratory Practice for Non-clinical Laboratory Studies). In particular, non-clinical laboratory studies for safety evaluation shall be conducted in a laboratory registered under Section 510 of the FDA act. Some of the requirements for these non-clinical laboratory studies are presented below:

- Testing facility involved in non-clinical laboratory studies should permit the FDA to inspect the facility and to inspect all records and specimens (CFR 58.15.a).

---

[15]FDA (2019). Recommended Content and Format of Non-Clinical Bench Performance Testing Information in Premarket Submissions. Retrieved from https://www.fda.gov/media/113230/download

- The study director should be appointed (CFR 58.33). The study director has overall responsibility for the study's technical conduct and the interpretation, analysis, documentation, and reporting of results and represents the single point of study control.
- A testing facility shall have a quality assurance unit (CFR 58.35).
- A testing facility shall have written standard operating procedures (CFR 58.81.a).
- Each study shall have an approved written protocol (CFR 58.120.a).
- All changes in or revisions of an approved protocol and the reasons, therefore, shall be documented, signed by the study director, dated, and maintained with the protocol (CFR 58.120.b).
- All raw data, documentation, protocols, final reports, and specimens (except those specimens obtained from mutagenicity tests and wet specimens of blood, urine, feces, and biological fluids) generated as a result of a non-clinical laboratory study shall be retained (CFR 58.190.a).

3.5.2 Clinical Studies

Clinical studies are essential for establishing the safety and effectiveness of the medical device. However, they can also be a very high hurdle for a manufacturer (time and money). That is why regulators take a risk-based approach in a request for clinical data. For example, in the US, clinical studies are most often conducted to support high-risk devices (PMA). Only a small percentage of 510(k) devices require clinical data to support the application.

The clinical validation requirements for PMA devices in the United States are outlined in.[16]

For most markets (with China as a notable exception), clinical data can come from clinical investigations conducted either domestically or in other countries.

Typically, clinical studies have to meet high ethical and quality standards, particularly in human subject protection. Regulators set

---

[16]FDA, PMA Clinical Studies, Retrieved from https://www.fda.gov/medical-devices/premarket-approval-pma/pma-clinical-studies

ethical and quality standards for their respective countries. However, in order for clinical data from a study conducted in a foreign country to be accepted in the submission, it must meet certain ethical and quality standards as well.

For example, the FDA requires that data from clinical investigations (conducted in or outside the United States) be from studies conducted in accordance with good clinical practice (GCP). These requirements are outlined in an FDA rule, "Human Subject Protection; Acceptance of Data from Clinical Investigations for Medical Devices."[17] GCP includes review and approval by an independent ethics committee and informed consent from subjects.

For investigations conducted in the US, the standards are even higher. In particular, the FDA requires applicants and sponsors to state whether the clinical investigation complied with 21 CFR parts 50 (informed consent and additional safeguards for children in research), 56 [Institutional Review Board (IRB) requirements], and 812 [Investigational Device Exemptions (IDE) requirements].

The requirements for outside the US (OUS) are outlined in 21 CFR 812.28. Suppose the OUS study cannot comply with GCP. In that case, the sponsor needs to either request a waiver or provide a "statement explaining the reason for not investigating in accordance with GCP and a description of steps taken to ensure that the data and results are credible and accurate and that the rights, safety, and well-being of subjects have been adequately protected."

In summary, clinical studies of medical devices must comply with FDA's human subject protection requirements (informed consent and additional safeguards for children in research) (21 CFR Part 50), Institutional Review Board (IRB) requirements (21 CFR Part 56), IDE requirements (21 CFR Part 812), Financial Disclosure for Clinical Investigators requirements (21 CFR Part 54) regulations. Requirements for *in vitro* diagnostics are covered in 21 CFR Part 809 (*In Vitro* Diagnostic Devices for Human Use).[18]

FDA has issued multiple other documents on clinical studies. The practically helpful guidance document for medical devices-

---

[17]FDA, Acceptance of Data from Clinical Investigations for Medical Devices. Retrieved from https://www.fda.gov/medical-devices/investigational-device-exemption-ide/acceptance-data-clinical-investigations-medical-devices
[18]FDA (2006). Information Sheet Guidance for IRBs, Clinical Investigators, and Sponsors. Retrieved from https://www.fda.gov/media/75381/download

related studies is "Information Sheet Guidance for IRBs, Clinical Investigators, and Sponsors."[19]

To address clinical trials during COVID-19 pandemic, FDA has issued an "FDA Guidance on Conduct of Clinical Trials of Medical Products during COVID-19 Public Health Emergency" document.[20]

3.5.2.1 Investigational device exemptions

A premarket submission may require clinical data. However, in order to use a medical device in a clinic, it needs to be legally marketed. How to solve this chicken-egg problem? Such a solution exists, and it is called an IDE.

An IDE allows the investigational (non-legally marketed) device to be used in a clinical study in order to collect safety and effectiveness data. "Clinical evaluation of devices that have not been cleared for marketing requires:

- an investigational plan approved by an IRB. If the study involves a significant risk device, the IDE must also be approved by the FDA;
- informed consent from all patients;
- labeling stating that the device is for investigational use only;
- monitoring of the study and;
- required records and reports."[21]

A clinical study under IDE has at least three major actors: a sponsor (manufacturer or importer), a clinical investigator, and an IRB from the institution, which employs the clinical investigator. The clinical investigator develops a protocol. The protocol needs to be reviewed by the IRB and for significant risk devices by the FDA. Thus, clinical study under IDE is limited to a health provider covered by an appropriate IRB. Exemption here will be multi-center studies, where multiple centers use the same protocol approved by a single IRB.

---

[19]FDA (2006). Information Sheet Guidance for Institutional Review Boards (IRBs), Clinical Investigators, and Sponsors. Retrieved from https://www.fda.gov/media/75381/download

[20]FDA (2020). Conduct of Clinical Trials of Medical Products During the COVID-19 Public Health Emergency. Retrieved from https://www.fda.gov/media/136238/download

[21]FDA, Investigational Device Exemption (IDE). Retrieved from https://www.fda.gov/medical-devices/how-study-and-market-your-device/investigational-device-exemption-ide

In the United States, IRBs have a high level of autonomy. In most cases, the IRB can make a significant risk/non-significant risk determination. And only significant risk devices require IDE approval by the FDA. In other jurisdictions, e.g., Canada, where it is called Investigational Testing Authorization (ITA), the process is more deterministic. In particular, Class I devices are exempt from ITA (however, they still need Research Ethics Board approvals). Class II–IV devices require ITA approval by Health Canada.

According to 21 CFR Part 812, there are three types of IDE studies: significant risk (SR) device studies, non-significant risk (NSR) device studies, and exempt studies.

Significant Risk Studies

An SR device means an investigational device that:

- "Is intended as an implant and presents a potential for serious risk to the health, safety, or welfare of a subject;
- Is purported or represented to be for a use in supporting or sustaining human life and presents a potential for serious risk to the health, safety, or welfare of a subject;
- Is for a use of substantial importance in diagnosing, curing, mitigating, or treating disease, or otherwise preventing impairment of human health and presents a potential for serious risk to the health, safety, or welfare of a subject; or
- Otherwise presents a potential for serious risk to the health, safety, or welfare of a subject [21 CFR 812.3(m)]."

Manufacturers of investigational SR device studies must get an approved IDE from FDA before starting their study. However, IRB approval is also required. For significant risk studies, it makes sense to go to the FDA first. In this case, the IRB will consider the protocol already approved by the FDA, and there is no need for IRB to reconsider the protocol, which the FDA can amend.

IRBs do not need to make the SR or NSR determination if the FDA has already made the risk determination.

Non-significant Risk Studies

An NSR device is an investigational device that does not meet a SR device's definition. If an IRB finds that an investigational medical device study poses a non-significant risk, the sponsor does not need to submit an IDE to FDA before starting the study.

Exempt Studies

In accordance with 21 CFR 812.2(b), sponsors and investigators of certain studies are exempt from the requirements of 21 CFR Part 812. Examples of exempt studies are consumer preference testing, testing of a device modification, or testing of two or more devices in commercial distribution if the testing does not collect safety or effectiveness data, or put subjects at risk.[22]

Studies of an already cleared [510(k)] or approved (PMA) medical device in which the device is used or investigated in accordance with the indications in the cleared/approved labeling are exempt. However, studies of a cleared device for a new use must comply with the human subject protection (informed consent and additional safeguards for children in research), IRB, and IDE regulations.[23]

In this chapter, we briefly touched on the regulatory aspects of clinical studies. Practical aspects of clinical studies will be discussed in Chapter 4, Product Development Process.

3.5.3 Human Factors Engineering (HFE)

The specialized testing facility will perform basic usability testing according to IEC 60601-1-6 (Medical electrical equipment—Part 1–6: General requirements for basic safety and essential performance—Collateral standard: Usability). However, regulators have a broader view of usability. They understand that user errors are among potential hazards, which the device can pose. Also, they understand that it is challenging to foresee user errors. Thus, to ensure medical devices' safety and efficiency, regulators mandate to integrate human factor engineering (HFE) into risk management.

HFE is a prerequisite for obtaining clearance to market Class II or III medical devices in the United States. Similarly, in Europe,

---

[22]FDA (2006), Information Sheet Guidance for Institutional Review Boards (IRBs), Clinical Investigators, and Sponsors. Retrieved from https://www.fda.gov/media/75381/download

[23]Wynbrandt, J (2016). Post-Approval Studies: Similarities and Differences from pivotal studies. Retrieved from https://www.imarcresearch.com/blog/post-approval-studies-similarities-and-differences-from-pivotal-studies

manufacturers need to self-declare that their medical products conform to IEC's HFE standard to qualify for a CE Mark. In HFE regulations, regulators are primarily concerned about the device's safety and effectiveness for the intended users in the use environment. The goal is to ensure that the user interface (UI) eliminates or reduces to the extent possible any use errors, which occur during the use of the device, that could cause harm or degrade medical treatment.[24]

FDA published a guidance document[25] on human factors and usability engineering. In most other jurisdictions, the IEC 62366-1 (Medical devices—Part 1: Application of usability engineering to medical devices.) standard is used. FDA guidance and IEC 62366-1 standard mandate that the manufacturer needs to apply a systematic process to analyze, specify, develop, and evaluate a medical device's usability related to safety.

As we mentioned before, HFE needs to be integrated with design controls and risk management. In this case, user errors can be considered among other hazards, and a standard risk management approach can be applied. However, there are certain specific expectations pertinent to HFE. In particular, it is expected that:

1. Design requirements and goals will be based on research into the intended users' needs and preferences.
2. Established UI design principles to hardware, software, and documents (e.g., printed and electronic instructions) will be applied.
3. Extensive testing will be conducted to determine if users can perform tasks safely and effectively.
4. Design enhancements will be developed to address the shortcomings revealed by testing.

We will examine these steps briefly.

3.5.3.1 Analysis and evaluation

The primary part of the HFE is a user research and user testing. There are several types of testing, which need to be performed at various product development stages.

---

[24]FDA (2016). Applying Human Factors and Usability Engineering to Medical Devices. Retrieved from https://www.fda.gov/media/80481/download
[25]FDA (2016). Applying Human Factors and Usability Engineering to Medical Devices. Retrieved from https://www.fda.gov/media/80481/download

Preliminary analyses and evaluations are performed to identify user tasks, UI components, and use issues early in the design process.

To identify critical tasks contextual inquiries, interviews, and formative evaluations are used.

Formative evaluations are also used to inform device UI design while it is in development. Primary methods of formative assessment (user research) are cognitive walk-through and simulated use tests.

In particular, the manufacturer needs to demonstrate that it conducted a proper level of user research.

What is the appropriate level? There is no correct amount or a precise recipe. However, any fresh feedback enhances the in-house body of knowledge.

Where to conduct user research? It is sensible to conduct research in primary and special (i.e., unique) markets. It is OK to generalize to other markets.

How to conduct user research? You can mix realistic user testing in medical establishments, remote interviews, and marketing and training specialists' in-house expertise.

3.5.3.2 Elimination or reduction of use-related hazards

Multiple user-related risk controls can be designed and implemented. However, as with other risk controls ISO 14971 prioritizes risk control options in order of preference and effectiveness:

- Inherent safety by design
- Protective measures in the actual medical device and/or manufacturing process
- Information for safety, such as labeling and instructions for use

Thus, the labeling and instructions for use should not be overused and treated as a last resort.

3.5.3.3 Human factors validation testing

In the medical device industry, the final (i.e., validation) usability test should provide evidence that the product effectively protects against use errors that could cause significant harm.

What kind of proof is needed? Manufacturers need data showing that most, if not all, test participants completed key tasks and that the residual risk associated with any use errors (i.e., mistakes) that occurred is negligible.

What is the size of the testing? FDA guidance specifies "a sample of 15 people to detect most of the problems in a UI constitutes a practical minimum number of participants for human factors validation testing." Note that if the device has more than one distinct population of users, then the validation testing should include at least 15 participants from each user population.[26] By comparison, IEC's standard does not mandate a specific number of test participants but does provide guidance on selecting an appropriate sample size.

Key Takeaways:

- HFE is a prerequisite for obtaining a clearance/approval to market medical devices in most cases and needs to be integrated into risk management.
- The FDA guidance document on HFE and IEC 62366-1 standard are not the same but have significant overlapping. It does make sense to design an HFE system, which will be compatible with all your primary markets.

---

[26]FDA (2016). Applying Human Factors and Usability Engineering to Medical Devices. Retrieved from https://www.fda.gov/media/80481/download

Chapter 4

Product Development Process

Any medical device is a product. Thus, in addition to regulatory aspects, we need to look at developing it as a product. It will be the goal of this chapter. We will start our discussion with general product development concepts. We will then discuss how to incorporate our knowledge of the regulatory environment and necessary prerequisites into product development.

The product development process (PDP) is a systematic approach to product development. It is a reasonably new concept. While companies develop products for a long time, product development's conceptualization is a much newer notion. Similarly, the widespread of product managers is quite a recent event. If in the 80s and 90s, only big companies had designated product managers, now it is mainstream, which is probably can be attributed primarily to the proliferation of startups.

There is no one size fits all solution for product development. There are multiple approaches to PDP, and various companies adopt different methodologies. Thus, we will consider several approaches to product development.

Project management is a concept, which is very close to product development. The development of a new product can be considered as a project. To some extent, product management evolved from project management. Thus, we will start with a brief overview of several project management approaches.

Bringing a Medical Device to the Market: A Scientist's Perspective
Gennadi Saiko
Copyright © 2022 Jenny Stanford Publishing Pte. Ltd.
ISBN 978-981-4968-25-6 (Hardcover), 978-1-003-31221-5 (eBook)
www.jennystanford.com

Waterfall Model

The waterfall model is probably the most well-known project management methodology. It is essentially a sequential process marked by steps that have to be complete in their entirety before moving to the next phase. It contains several distinct stages (e.g., planning, design, development, testing). The next phase starts after the successful completion of the previous phase. The typical waterfall model is depicted in Fig. 4.1.

Waterfall-based models are well-established methodologies widely accepted in many industries, from construction to consumer goods to space technologies. In particular, it is widely adopted in the medical device industry. The reason behind its popularity is that it is well defined, and all regulatory requirements can be easily integrated into this process.

However, this model has several significant drawbacks. One of the primary reasons is the cost of errors. If an error were introduced at an early stage (e.g., planning), it would most likely be identified only at the late stage (e.g., testing). Thus, the error propagates, and its overall cost of this error can be pretty significant. It is not just the cost of rework. It may also include the rework of other parts, which design was impacted by an erroneous decision. Thus, it may have a snowball effect. The earlier the error was introduced, the higher impact on the overall cost. That is the primary reason why most of the projects go over time and over budget (according to industry estimates, more than 70% of all projects go over time and budget).

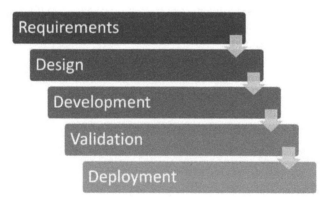

Figure 4.1   Waterfall model.

For example, according to the United States Federal Government estimates, 94% of IT projects commissioned by the government were over budget and behind schedule. 40% of them have never been delivered.

The other drawback is that the waterfall model is not adaptive to changes. Any change to the original requirements is typically managed through change requests, increasing the cost and pushing the project's schedule.

Iterative Models

An iterative process is an alternative to the pure sequential product development process. Within this approach, the product is developed in cycles. Each cycle includes all or almost all product development phases (design, development, testing) but on a much smaller scale. During one iteration (cycle), the product or a part of the product is being designed, developed, and tested (including integration with the product). The next cycle (or iteration) gradually increases complexity, etc. So, it is like we start with a skeleton and progressively add pieces of flesh to it. The typical iterative (Agile) model is depicted in Fig. 4.2.

The primary advantage of such an approach is its adaptability to changes. New requirements are not just permitted but even welcome. Any erroneous decision most likely will be identified during regular user demonstration and will not propagate into the final design.

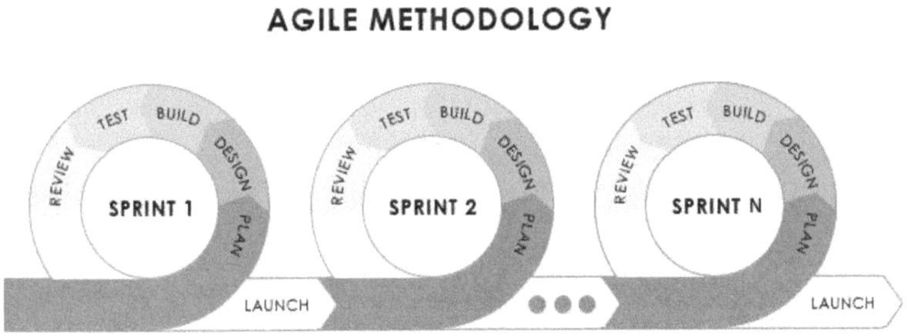

AGILE METHODOLOGY

Figure 4.2   Iterative (Agile) model.

There are multiple iterative frameworks available. Kanban and Scrum are probably the most well known among them. They are different in details and rituals; however, the methodology's core (incremental, iterative approach) is the same. This family of iterative approaches is commonly known as Agile methodologies. Most Agile development methodologies split product development work into small increments. Iterations, or Sprints, are short time periods that typically last from one to four weeks. This approach minimizes overall risk and allows the product to adapt to changes quickly by minimizing upfront planning and design.

One of the Agile methodology's key elements is the high involvement of a customer in the product development process. The motto of Agile methodologies is to design with a user, not for a user. The deliverables of iterations are demonstrated to customers, and customers' feedback is immediately integrated into the product.

The Agile methodology was popularized by an Agile Manifesto,[1] which was developed and signed in 2001. Agile methodologies are widely adopted in software development across multiple industries. There are known issues with adaptations of Agile approaches in large organizations with legacy infrastructure. But, they are popular and successful in a startup environment.

However, the iterative approach was not invented in 2001. There were multiple precursors of the Agile methodology. The origins of the iterative methods can be traced back at least to the 1930s when Walter Shewhart of Bell Labs began applying Plan-Do-Study-Act (PDSA) cycles.[2] Quality guru Edwards Deming used his methodology after World War II in Japan, where he trained Toyota's managers, ultimately developing the famous Toyota Production System. Iterative and incremental development methods were used to design the X-15 hypersonic jet in the 1950s. Iterative approaches began mushrooming in the 1990s in various industries. The next important milestone was 1995 when Jeff Sutherland and Ken Schwaber presented the scrum process at Texas's conference.

Thus, while Agile Manifesto was initially developed for software development, the iterative approach has multiple roots and can be adapted to other fields, including hardware development.

---

[1]Beck K, Grenning J, Martin RC, et al. (2001). "Manifesto for Agile Software Development". Agile Alliance. Retrieved from https://agilemanifesto.org/

[2]Rigby DK, Sutherland J, and Takeuchi H (2016). The Secret History of Agile Innovation, *HBR*. Retrieved from https://hbr.org/2016/04/the-secret-history-of-agile-innovation,

Agile software development methods were initially seen as best suitable for non-critical product developments and not suitable for regulated industries, such as medical devices. However, in recent years there are multiple successful adaptations in regulated industries, including MedDev.

4.1   Product Development Methodologies

The waterfall model is a generic project management methodology. It applies to a variety of projects, from construction to IT to space technologies. However, it applies to product management as well. In particular, each product can be considered as a project with a set of distinct phases. Thus, a good adaptation of the waterfall model to the product development needs to specify these stages.

4.1.1   Stage-Gate Model

A Stage-Gate® model is a waterfall-based technique in which a product development process is divided into stages separated by gates. At each gate, the development process's continuation is decided by (typically) a manager or a steering committee. The decision is based on the information available at the time, including, e.g., business case, risk analysis, availability of necessary resources (money, people with correct competencies), etc.

The Stage-Gate model was developed by Robert G. Cooper[3] from McMaster University.

The traditional Stage-Gate process has five stages (plus one preliminary) and five gates. The stages are:[4]

- Discovery (Stage 0)
    - The discovery stage is the first part of any product development, whether or not the Stage-Gate model is being utilized. It is a pre-work designed to uncover opportunities and generate new product ideas.

---

[3]Cooper R.G., *Winning at New Products: Creating Value Through Innovation*, 4th edition. New York, NY: Basic Books (division of Perseus Books), 2011.
[4]There are multiple sources of this description. This description is loosely adapted from phase-gate process. Retrieved from Wikipedia. https://en.wikipedia.org/wiki/Phase-gate_process.

- Scoping (Stage 1)
  - During this step the main goal is to evaluate the product and its corresponding market. The researchers must recognize the product's strengths and weaknesses and what it is going to offer to the potential consumer.
- Build Business Case (Stage 2)
  - This stage is the last stage of concept development where it is crucial for companies to perform a solid analysis before they begin developing the product.
- Development (Stage 3)
  - During the development phase of the Stage-Gate process, plans from previous steps are actually executed. The product's design and development is carried out, including some early, simple tests of the product and perhaps some early customer testing. The product's marketing and production plans are also developed during this stage.
- Testing and Validation (Stage 4)
  - The purpose of this stage is to provide validation for the entire project. The areas that will be evaluated include: the product itself, the production/manufacturing process, customer acceptance, and the financial merit of the project.
- Launch (Stage 5)
  - The product launch is the culmination of the product having met the proper requirements of the previous stage-gates. Development teams must come up with a marketing strategy to generate customer demand for the product.
  - The company must also have to decide how large scale they anticipate the market for a new product to be and thus determine the size of their starting volume production.
  - In addition, part of the launch stage is training sales personnel and support personnel who are very familiar with the product and can assist in market sales.

There are multiple variations of the Stage-Gate process. A typical structure for the Stage-Gate process with 4+1 stages and its main elements are depicted in Fig. 4.3.

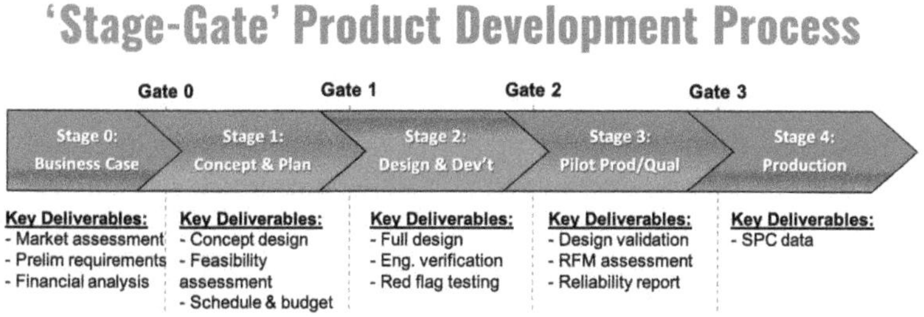

'Stage-Gate' Product Development Process

Figure 4.3 Stage-Gate PDP. Reproduced with permission from footnote 5.

Stages have a common structure and consist of three main elements:

1. Activities
   - Activities consist mainly of information gathering by the project team to reduce key project uncertainties and risks.
2. Integrated Analysis
   - The project team undertakes an integrated analysis of the results of the activities.
3. Deliverables
   - Deliverables of stages are the results of integrated analysis and these are the input to the next Gate.

Preceding each stage is a gate or a Go/Kill decision point. A decision at the gate is based on the previous stage's outcomes and predetermined criteria, which can be split into Must Meet and Should Meet. The output at the gate is a decision (Go/Kill/Hold/Recycle).

According to footnote 6, "Stage-Gate is used by producers of physical products—both consumer goods (such as Procter & Gamble and General Mills) and industrial goods (such as 3M, DuPont, ITTIndustries, BASF, Siemens and Emerson Electric)—and also service providers (such as banks and telephone companies)."

4.1.2 Agile Methodologies

On an iterative methodologies side, there are multiple Agile product development methodologies.

---

[5]Giltner D, Turning Science, Product Development Process, OSA webinar, Jun 24, 2020.
[6]Cooper RG, The Stage-Gate®System: A Road Map from Idea to Launch – An Intro & Summary. Retrieved from http://www.bobcooper.ca/images/files/articles/2/1-The-Stage-Gate-System-A-Roadmap-From-Idea-to-Launch.pdf

Actually, "Agile" and "Sprint" are buzzwords into product development. Most likely, if you hear about product development, nowadays, it mainly refers to Agile.

However, there are good reasons for such popularity. Agile facilitates the delivery of a customer-centric product. It improves odds for adoption by users.

Following Agile, the product is being developed iteratively, in small increments. Each iteration (Sprint) is typically two weeks long. However, two weeks duration is a standard for software development. Nowadays, the utility of Agile has been proven beyond software development. However, some changes to the process are required. For hardware development, a two-week Sprint duration is most likely impractical (e.g., to accommodate parts purchase and delivery). Five to seven-week Sprints can be more realistic.

The Sprint consists of a Sprint planning meeting, daily Scrums, and a final Sprint meeting, where deliverables are presented.

During the Sprint planning meeting, team members select tasks for the following Sprint from the Backlog, the ledger for all product ideas. One way to control the Sprint scope is to assign a numeric score to each task's complexity (e.g., 1 for easy, 3 for medium, 5 for hard). Then, the overall complexity of all tasks selected for the Sprint should be close to a predetermined level (e.g., 30), which is based on the team's size and can be found experimentally.

One common approach is to consider everyone in the scrum team with an equal voice and allow scoring it. QA has the same voice as a lead developer, and then as a group, they decide on the score for the "story." This is an essential part as some tasks can be 1 on developmental effort but 8 on QA. And in reverse. Another common way is to allow a fixed number of scores, for example, 1, 3, 5, 8, and 13. If the 13 is given, it means the story is too big; thus, it has to be broken down into components and scored individually. Anytime someone puts 13 on a story, they have to defend it, which may turn into a good discussion. It is up to the group to decide whether to split it or take it as is (e.g., using the medium score).

During daily Scrums (or stand-ups), team members report their progress for the previous day and plans for the coming day.

The final Sprint meeting is a demo, where the solution developed during the Sprint is demonstrated to stakeholders. Oftentimes, the final Sprint meeting also includes a retrospective component (look back on the Sprint).

Sprints are facilitated by Scrum masters, whose role is different from project managers. While project managers manage the budget, scope, and resources, the Scrum master primarily is a coach and facilitator for the product team.

A product owner is the business representative on the scrum team. One of his/her primary responsibilities is to write "stories." The product owner is a Scrum role, which a product manager typically plays. However, in some organizations, the product owner may become a separate job.

In addition to regular product development Sprints, the product development process starts with a Discovery Sprint. During the Discovery Sprint, the product team works on understanding the user problem.

All product ideas are managed through the Backlog.

I would like to finish the brief discussion of Agile with some practical notes. The primary reason why the iterative approach offers better results (in theory at least) is that there is more context, and developers can make better decisions close to the user. So, ideally, actual users should be involved. However, if it is not possible, then "personas" are used. Personas are archetypes of key users that are associated with specific features or workflows in a product. These are important as a customer (purchaser), and a user may be different. One could be a middle-aged college-educated professional, the other one 70+ person. This will, of course, drive the type of behavior for a feature. It is good to develop these ahead and associate them with workflows and features. In the MedDev context, a medical doctor, nurse, and patient will use different features and workflows. Those are all different personas, and they absolutely should be considered when building a UI.

4.2 Integration of Product Development Process with Regulatory Requirements

In regulated industries, product development should be aligned with regulatory requirements. That is why the company's product development methodology should be amended accordingly.

4.2.1   Waterfall Models in MedDev PDP

Waterfall-based product development models are very well suited for regulated industries like MedDev. In particular, regulatory requirements can be easily integrated into waterfall-based product development models.

Any product development process starts with a planning phase. During this phase, the risks are identified and assessed in parallel with functional requirements, and risk management plans are being developed. In addition to that, design and development planning is being performed.

During the design and development phase, the design team converts design inputs into design outputs. During this phase, Design Reviews take place.

Finally, the design is being validated and verified during the testing phase. During this phase, the Risk Management Report has to be created, and the risk management review has been conducted.

4.2.2   Agile Methodologies in MedDev PDP

Is it possible to tailor an iterative adaptive approach to the regulated environment? The short answer is "Yes." For example, the FDA has accepted the use of Agile software development since 2013, specifically for medical device software. The standards do not prescribe a particular product development methodology. Instead, they define that certain key points that need to be addressed. These key points are:

- Quality assurance (QA): Systematic and inherent quality management underpinning a controlled process and the product's reliability and correctness.
- Safety and security: Formal planning and risk management to mitigate safety risks for users and securely protecting users from unintentional and malicious misuse.
- Traceability: Documentation providing auditable evidence of regulatory compliance and facilitating traceability and investigation of problems.
- Verification and Validation (V&V): The assurance that the product is designed according to specs and meets user requirements.

Thus, the goal is to integrate these requirements into the fabric of iterations of Sprints.

It can be done through embedding formal steps and documents (e.g., user requirements specification, functional specification, design specification, code review, unit tests, integration tests, system tests) into the product development process. For example, one way to do it is through test-driven development, TDD. In this approach, unit tests are developed first and then built towards them.

In addition to that, the Sprint planning meetings and Sprint final meeting need to be amended accordingly. From the risk management perspective, Sprint planning meetings need to include risk identification and evaluation. Similarly, the final Sprint meeting needs to include residual risks assessment and discussion.

Design verification can be conducted as a part of the testing during the Sprint. Design validation is being performed during the final Sprint meeting.

4.2.3   MedDev PDP: Waterfall or Agile?

Both waterfall-based and Agile methodologies have multiple strong proponents in the MedDev. However, multiple FDA-regulated companies' analysis shows that most MedDev companies use the waterfall-based methodologies (predominantly Stage-Gate). It is particularly true for market leaders. There is undoubtedly an inherited bias in these conclusions. Most market leaders are medium and large companies; thus, this analysis is biased toward medium and large companies. However, the real reason for that is probably still regulatory requirements. With multiple interdependences on regulators and auditors (e.g., notified bodies), it is easier to stick with waterfall-based methodologies and clearly-defined phases. The other probable reason for that is human factors. Implementing regulatory compliance (e.g., good manufacturing practices, GMP) often involves industry professionals who are most familiar with traditional, proven methodologies.

However, there is still a simple (and compliant!) way to speed up the product development process. It can be done by splitting the project into phases and using waterfall models (namely, Stage-Gate) for high-level PDP and Agile methodologies within each phase.

Such a hierarchic approach may address waterfall models' primary challenges while maintaining the rigid structure and regulatory oversight.

4.3 PDP Phases

This section will follow the industry-standard separation of the product development process into several distinct phases. The typical split is R&D, product development, (pre-clinical), clinical, and regulatory phases. Note that the pre-clinical phase is relatively short and can be combined with the product development phase.

It should be mentioned that such separation is typical for Class III devices. For Class I and Class II devices, the clinical phase maybe not necessary.

4.3.1 R&D and Product Development

Integration of regulatory requirements into the product development process brings a lot of complexity to the development process. How to navigate it, especially early in the product development process? At this stage, in many cases, the product is very far from the final look and feel and requires many changes before going to market.

The short answer is that you can split the product development lifecycle into several distinct phases (e.g., prototype development and final product development) and apply different approaches to these phases.

The regulators are concerned with the safety and efficiency of the final product. Thus, the final product development needs to include the quality management system and risk management requirements. However, it does not preclude you from experimentation and pivoting during the prototype development phase. It can be excluded entirely from the regulatory requirements.

Thus, if you split the product development lifecycle into several phases, you are safe from a compliance perspective. And it will not impede you from rapid changes in the direction of the early stages of the product development. However, it will be beneficial to document a transition from an "unregulated" phase to a "regulated" product development phase.

4.3.2 Clinical Phase

The clinical phase can be one of the most challenging parts of the MedDev product development. That is why it is of paramount importance to understand regulatory requirements before starting it.

Firstly, as we already mentioned in many cases, clinical validation is not required. Most Class I devices are exempt. For non-exempt devices, just a small fraction (around 10%) of 510(k) submissions requires clinical validations. And only Class III devices require it.

Secondly, even if clinical validation is required, the level of its complexity varies. There are two main clinical evaluation types, an observational study and a randomized control trial (RCT).

Under an observational study, a change in a patient's clinical pathway is made, and measurements are taken before, after, and at some points during this change. The disadvantage of this method is that differences in outcomes can be affected by some other factors.

The design of clinical trials aims to reduce this bias. The RCT has two or more patient groups (arms), one for whom the patient care pathway is changed (experimental arm), and the other where there is no change (control arm). The intervention's efficacy (i.e., the device's use) is measured through the differing outcomes between the two groups. The disadvantage of this method is the larger number of patients involved (double compared with an observational study due to the presence of two arms), which inevitably leads to operational complexities (e.g., recruitment) and significant cost increase.

While a randomized clinical trial may be required in some instances, in many cases, a clinical study (for example, an observational study with a single arm) will be sufficient.

In drug approval, there are up to five phases of clinical trials: Phase 0, Phase I, Phase II, Phase III, and Phase IV. Each phase is considered a separate trial. After completion of a phase, investigators need to submit their data for approval from the FDA before continuing to the next phase. As a rule of thumb, the number of participants increases 3–10 fold at each phase. Phase 0 trials aim to learn how a drug is processed in the body and how it affects the body and include 10–15 participants. Phase I trials typically include 15–30 participants and aim to find the best dose of a new drug with the fewest side effects. Phase II trials further assess the safety and if a drug works and typically include from 30 to 100 participants. Phase III Phase III trials compare a new drug to the standard-of-care drug and include hundreds or thousands of participants. For example, Phase III trials for COVID-19 vaccines included 5,000–20,000 patients. Phase IV typically comprises significantly more than 300 participants.

Phase III clinical trials are often needed before the FDA will approve the use of a new drug for the general public. Phase IV trial is typically conducted on the approved drug.

4.3.2.1   Regulatory requirements for MedDev

For Class III medical devices, clinical validation is typically done through two stages of clinical trials: a pilot study and a pivotal study.

The focus of the pilot study is safety and feasibility. It is run on a small number of patients.

Pivotal study is similar to Phase III clinical trial in drug development. It is intended to demonstrate and confirm the safety and efficacy of the medical device or clinical diagnostic procedure and estimate the incidence of common adverse effects.

The pivotal study is run on hundreds or thousands of patients. The useful rule of thumb here is Hanley's "Rule of 3." "The probability of adverse and undesirable events during and after operations that have not yet occurred in a finite number of patients (n) can be estimated with Hanley's simple formula, which gives the upper limit of the 95% confidence interval of the probability of such an event: upper limit of 95% confidence interval = maximum risk = 3/n (for n > 30)."[7]

In an analysis of pivotal trials on medical devices conducted between 2006 and 2013, the median duration was three years, with another two years needed for FDA review and approval for marketing.[8] Trials had a median enrollment of 297 patients.

Also, note that the regulatory requirements vary between different branches of the FDA:

Medical devices regulated by CDRH (Class III)

- A pilot study may be necessary to demonstrate proof-of-concept/feasibility

---

[7]Eypasch E, Lefering R, Kum CK, Troidl H (1995). Probability of adverse events that have not yet occurred: a statistical reminder. *BMJ*, 311(7005): 619–620.

[8]Rising JP, Moscovitch B (2015). Characteristics of pivotal trials and FDA review of innovative devices. *PLoS ONE*, 10(2): e0117235.

- A pivotal study is required to demonstrate safety and effectiveness. These studies can be comparative or single arm. They also may use historical data for comparison.

Biologics regulated by CBER

- Two adequate and well-controlled clinical studies.

4.3.2.2   Clinical trial design considerations

If you need to go through clinical trials, there are several ways to make it a little bit simpler.

Firstly, the proliferation of Contract Research Organizations (CROs) has changed the clinical trial landscape dramatically. Nowadays, a significant part of clinical work can be outsourced to CROs. You don't need to have study coordinators in-house, search for clinical partners, or recruit patients. All these functions can be done by the CRO, bringing down the cost and complexity of clinical work significantly.

Secondly, working with a patient group (or patient advocacy group) can significantly simplify clinical work. Patient groups are very active organizations. In many cases, they built the whole ecosystem around them. In particular, they have a network of patients, who are willing to participate in trials, and they work with doctors who can run these studies. Moreover, in some cases, access to the patient network can be an in-kind contribution, significantly bringing down the cost.

During the design of clinical validations, it is important to take into account several other considerations:

- Who is the primary user of the device? For example, if the device is set to be used in primary care, then this is where your clinical evaluation should be. The device's efficacy can be very different when used in different settings, for example, by a specialist or in primary care.
- A manufacturer may want to measure different parameters for different purposes, i.e., a device may improve healing speed. However, it might also improve the patients' quality of life—this may be just as important both for manufacturer and patient. Multiple measurements can be built into a single

trial, and this will almost always be more cost-effective than running multiple trials.

- While designing a clinical trial, it may also be beneficial to include health-economics factors. It may help in the future with reimbursement approval. Some differences between requirements for "regulatory approval" and "reimbursement approval" will be outlined in Chapter 8 (Business Model).

4.3.3   Regulatory Phase

During this phase, the product development has to be completed and stabilized. The team needs to complete all verifications and validations, including bench testing, pre-clinical testing, clinical testing, EMC/Safety testing, usability testing, etc.

At this stage, the regulatory pathway needs to be well understood and all necessary preparation completed. It is probably OK to address minor gaps; however, all required systems (e.g., Quality Management System and risk management) need to be implemented beforehand.

As a result, all documentation required for submission needs to be prepared, packaged, and submitted to a regulatory body. Depending on the medical device's class and regulator, the amount of documents required for submission varies considerably.

The regulatory submission is an essential milestone of the project. Depending on the regulatory pathway, the clearance/approval may take from several months to several years. While in some cases, there are certain time limits set by overseeing legislative bodies (as in the case of FDA), the regulatory clock stops with any information request or question raised by the regulator. In this case, the duration of the overall process depends on the availability of the information.

It raises a strong case for pre-submission communications with regulators. For example, the FDA has numerous formal or informal pre-submission communications mechanisms discussed briefly in Chapter 2 (Regulatory Environment). Aligning with FDA beforehand, which can be done during earlier phases, may save a significant amount of time and money during the regulatory phase. For example, it is highly beneficial to solicit the FDA's opinion on clinical study design.

4.4   Other Helpful Considerations

4.4.1   Prototyping

One of the primary reasons for product failures is the lack of product/market fit. This means there is a mismatch between the problem in the market and the solution that is provided. The worst case of this is when the product doesn't solve an actual problem. Product/market fit can be researched but always needs to be validated by engaging users in the design process. Prototyping is a primary tool for doing it. The key point of prototyping is to get the product tested by users and the users' feedback incorporated into the design. Fortunately, current technologies (e.g., 3D-printing) allow rapid prototyping.

There are multiple approaches to prototyping. It is also typically an iterative process. One can start with a low-fidelity prototype, which can be as simple as sequences of paper drawings. Non-critical detail and the visual appeal are purposefully foregone in favor of the functionality of a feature. It is critical to observe the user's natural behavior and not interrupt as it will provide unbiased feedback. Developers and designers will have their own opinion of how it should be used, but it is the user that will dictate how those features are discovered and used.

Once the prototype is tested, the next generations of prototypes (high fidelity) can be designed and developed. For software products, it can be mock apps and websites; for hardware products, it can be 3D-printed.

4.4.2   Customer Development

Together with business model design and Agile engineering, customer development forms three pillars of a Lean Startup methodology. It focuses on continual product iteration and refinement based on customer feedback.

Steve Blank developed and popularized customer development methodology. Customer-centric development and the customer-centric company are the core of this philosophy, radically different from a traditional product-centric approach.

Customer development is an empirical approach and requires a lot of experimentation, including customer interviews. The probably most notable quote regarding customer development is "There are no facts inside your building, so get outside."[9]

According to Agile Alliance, the customer development framework consists of four steps:

- "Customer discovery—Understand customers and their needs that you may be able to satisfy.
- Customer validation—You have a product that will satisfy your customer's needs.
- Company creation—You determine whether your product will satisfy all the customers' needs.
- Company building—You can grow your organization in order to support the demand for your product."[10]

All these steps need to be executed at various stages of the company/product development.

You have to start with the customer discovery at very early stages, preferably even before the company creation. Who are your customers? Which problems do they have? Do you have the business case in the first place? It needs to be verified by experimentation. It helps to set the course of the venture and avoid costly mistakes in the future.

During product development, the product needs to be developed with customers. It is the focus of the customer validation step.

The primary benefits to customer development are that it provides a team with an approach to validating assumptions and ensuring that the team delivers the right thing.

As we mentioned before, customer development is the Lean Startup concept's pillar and is deeply integrated with the business model development. To guide the business model development, the Lean Canvas framework is particularly useful. It will be described briefly in Chapter 8 (Business Model).

4.4.3 MVP

The goal of prototyping and customer development is to establish the minimum viable product (MVP). As it quite clear from its name, MVP

---

[9]Blank S, Dorf B, The Startup Owner's Manual: The Step-By-Step Guide for Building a Great Company, Pescadero, CA , K & S Ranch (2012).

[10]Agile Alliance, Customer Development. Retrieved from https://www.agilealliance. org/glossary/customer-development

is a minimalistic set of features, which are essential for the customer. The purpose of MVP is to solicit early customers' feedback to guide future product development and avoid lengthy and unnecessary work.

MVP is a core artifact in an iterative process of idea generation, prototyping, presentation, data collection, analysis, and learning. The process is iterated until a desirable product/market fit is obtained or until the product is deemed non-viable.

MVP needs to be tested with customers to determine whether they are willing to pay for it. It means that MVP needs to be an industry-grade solution. You can make a cheap and ugly prototype; however, MVP needs to have a look and feel of the real product.

A common issue with MVP is the incorrect scope of the problem it is solving. Ensure you understand the minimal problem that your problem is solving. If it does not solve a problem, even the smallest one, it is not an MVP. On the other hand, if you have packed many features into it, it is most likely too big in scope, and you have missed many points you could have iterated on.

4.4.4 Pivoting

In addition to scientific research, the scientific method can be applied to virtually any aspect of startups, including product development. This approach was popularized by Steven Blank and Eric Reis in the Lean Startup methodology.

The scientific method consists of several steps. In step one, you formulate your hypothesis. In step two, you test your hypothesis. Then, you analyze your results. If the results confirm your hypothesis, then you stick to it or slightly modify it. However, if they contravene your hypothesis, you may need to discard it and go back to the drawing board to design a new one.

The same may happen to your product. You came up with your hypothesis that your device solves a specific clinical problem. Thus, the next step is to test it with users. If the user feedback confirms your hypothesis, it is excellent. You are on the right track. However, in some cases, the feedback can be mixed or even negative. It is also great. It may save you a lot of time and money. However, you may need to revisit your product or some aspects of it (e.g., business model).

Sometimes, you may even need to rethink your strategy entirely and implement radical changes to the original plan. This process is called a pivot.

Pivoting is a primary method of how startups evolve. The classic example here is Raytheon, a major defense contractor, and missile manufacturer. The company started in 1922 with a focus on refrigeration technology. The company's first product was a gaseous (helium) rectifier.

Viagra is a famous medical example. Pfizer originally developed the sildenafil compound to treat high blood pressure (hypertension) and chest pain due to heart disease (angina pectoris). Its useful side effect was discovered during the heart clinical trials.

From the startup world, Slack is a famous example these days. Initially, it was a video game company, and as they were running out of money, they did a random pivot on an internal tool they built for internal communications.

It may be almost impossible to make the product and your company right from the first attempt. You need to test and challenge every assumption about your product and business model.

Steven Blank defined a pivot as "changing (or even firing) the plan instead of the executive (the sales exec, marketing or even the CEO)."[11]

Some companies even make pivoting a part of their business strategy. They constantly pivot in different directions and observe what sticks to the wall.

4.4.5 TRL

Technology readiness levels (TRL) is a helpful (and widespread) conceptual framework and assessment tool used in product development. TRL is a method for estimating the maturity of technologies. While NASA originally designed it in the 1970s, it got significant traction with various government agencies. The US Department of Defense has used this methodology since the early 2000s. The European Commission uses it on EU-funded research and innovation projects since 2010. In particular, the EU Horizon

---

[11]Blank S, Dorf B, The Startup Owner's Manual: The Step-By-Step Guide for Building a Great Company, Pescadero, CA, K & S Ranch (2012).

2020 program uses it since 2014. Similarly, in Canada, it is used by Innovation Canada.

There are multiple versions of the TRL system adapted to various industries. Consequently, definitions of each level vary slightly between various agencies. We will use the Innovation Canada system[12] as a baseline. It is still generic but fits well with the MedDev market:

"**TRL 1**—Basic principles of concept are observed and reported
Scientific research begins to be translated into applied research and development. Activities might include paper studies of a technology's basic properties.

TRL 2—Technology concept and/or application formulated
Once basic principles are observed, practical applications can be invented. Activities are limited to analytic studies.

TRL 3—Analytical and experimental critical function and/or proof-of-concept
Active research and development is initiated. This includes analytical studies and/or laboratory studies. Activities might include components that are not yet integrated or representative.

TRL 4—Component and/or validation in a laboratory environment
Basic technological components are integrated to establish that they will work together. Activities include integration of "ad hoc" hardware in the laboratory.

TRL 5—Component and/or validation in a simulated environment
The basic technological components are integrated for testing in a simulated environment. Activities include laboratory integration of components.

TRL 6—System/subsystem model or prototype demonstration in a simulated environment
A model or prototype that represents a near desired configuration. Activities include testing in a simulated operational environment or laboratory.

TRL 7—Prototype ready for demonstration in an appropriate operational environment

---

[12]Industry Canada, Technology readiness levels. Retrieved from https://www.ic.gc.ca/eic/site/080.nsf/eng/00002.html

Prototype at planned operational level and is ready for demonstration in an operational environment. Activities include prototype field testing.

TRL 8—Actual technology completed and qualified through tests and demonstrations

Technology has been proven to work in its final form and under expected conditions. Activities include developmental testing and evaluation of whether it will meet operational requirements.

TRL 9—Actual technology proven through successful deployment in an operational setting

Actual application of the technology in its final form and under real-life conditions, such as those encountered in operational tests and evaluations. Activities include using the innovation under operational conditions.

The more MedDev-specific system was developed by the US Army Medical Research and Materiel Command (MRMC). Their TRL system is linked to "FDA events." The description of the MRMC system can be found in Appendix C. It can be particularly useful while dealing with the US granting agencies.

The TRL system is a base for various grant programs. In this case, grant agencies specify which TRL is eligible for a particular grant funding. For example, Innovation Canada considers only TRL 7–9 projects suitable for the Build in Canada Innovation Program.

In Fig. 4.4, one can see how technology maturity fits into product development process.

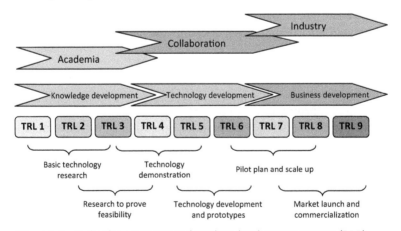

Figure 4.4 Technology maturity and product development process (PDP).

Stages 1–3 are considered very early stages for the technology. Grants are the only practical way of funding at this stage. Thus, the technology typically resides in academia.

Stages 4–7 are referred to as technology development. At this point, the technology is being transferred and developed in a company.

Finally, stages 7–9 can be referred to as a commercialization stage. At this stage, the product is being finalized and goes through clinical and regulatory phases.

Based on this, the product development process can be split into several distinct phases: Innovation Phase (TRL 4–6), Clinical Phase (TRL 7–8), and Regulatory Phase (TRL 9).

Chapter 5

Timelines and Capital

Now, we understand the scope of efforts required to get the device through the regulatory phase. Equipped with this knowledge, we can try to answer how much money we need to bring the medical device to the market? However, we will start with a question of how long it takes to get a device to the market.

5.1 How Long Does It Take?

The answer to the questions "How long does it take and how much money is required to commercialize a MedDev product?" is important from several perspectives. Firstly, it helps to set realistic expectations for the founding team. Secondly, it can be used for setting up realistic budgets for investors.

   The most comprehensive study on this topic conducted to date was the 2010 survey of 204 US MedDev companies[1]. The findings were quite startling, and we will discuss them below briefly. However, there are certain reservations about the applicability of that study. Firstly, it was done in 2010. Things might change significantly since then. For example, the least burdensome approach was mandated

---

[1]Makower J., Meer A., Denend L., FDA Impact on U.S. Medical Technology Innovation: A Survey of Over 200 Medical Technology Companies. November 2010. Retrieved from https://www.advamed.org/sites/default/files/resource/30_10_11_10_2010_Study_CAgenda_makowerreportfinal.pdf

Bringing a Medical Device to the Market: A Scientist's Perspective
Gennadi Saiko
Copyright © 2022 Jenny Stanford Publishing Pte. Ltd.
ISBN 978-981-4968-25-6 (Hardcover), 978-1-003-31221-5 (eBook)
www.jennystanford.com

by the 21st Century Cures Act, which indeed reduced the regulatory-associated costs. Secondly, the survey is probably tilted to large public companies (the study was limited to MedDev industry association members). The startups are known for making things faster and cheaper.

In addition to that study, there are several other sources of information. Starfish Medical, a product development boutique, researched the costs of bringing a medical device to market.[2] They used four methodologies to provide different angles on the product cost. In addition to mentioned above study, they analyzed public information, point-of-care commercialization case study,[3] and "the no design" approach to account for regulatory expenses.

The 2010's survey focused on several research questions. One of them was a comparison of the US and EU regulatory systems on premarket process duration. To make this comparison accurate, we need to account for the premarket review process's scope in both markets. While both systems require some assurance of safety, within the EU, a device manufacturer must prove only a device "functions as intended," whereas in the United States, they must demonstrate effectiveness.[4] However, the survey data still gives an important indication of realistic timelines for both markets.

5.1.1 510(k) Product

According to Makower et al.,[5] "Survey respondents reported that the premarket process for 510(k) pathway devices (of low-to-moderate-risk) took an average of 10 months from first filing to clearance. For those who spoke with the FDA about conducting a clinical study for their low-to-moderate risk device before making a regulatory

---

[2]Drlik M., Starfish Medical, "How Much Does It Cost to Develop a Medical Device?" Retrieved from https://starfishmedical.com/assets/StarFish-Whitepaper-Cost-to-Develop-Medical-Devices-July-2020.pdf

[3]Chin C. D., Linder V., Sia S. K. (2012). Commercialization of microfluidic point-of-care diagnostic devices, *Lab on a Chip*, **12**, 2118–2134.

[4]Maak T. G., Wylie J. D. (2016) Medical Device Regulation: A Comparison of the United States and the European Union. *The Journal of the American Academy of Orthopaedic Surgeons*, **24**(8), 537–543.

[5]Makower J., Meer A., Denend L., FDA Impact on U.S. Medical Technology Innovation: A Survey of Over 200 Medical Technology Companies. November 2010. Retrieved from https://www.advamed.org/sites/default/files/resource/30_10_11_10_2010_Study_CAgenda_makowerreportfinal.pdf

submission, the premarket process took an average of 31 months from first communication to being cleared to market the device. In contrast, respondents said it took them an average of 7 months in Europe from first communication to being able to market the same (or equivalent) device." These results are depicted in Fig. 5.1.

Figure 5.1 510(k) and CE Mark regulatory timelines. Reproduced with permission from footnote 6.

It took approximately 20 months to develop a product and around one year to develop a clinical unit (see Fig. 5.2). The clinical and regulatory phases took around 45 months.

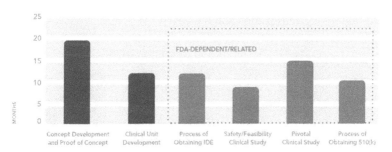

Figure 5.2 Average time by the stage for 510(k) product. Reproduced with permission from footnote 7.

---

[6]Makower J., Meer A., Denend L., FDA Impact on U.S. Medical Technology Innovation: A Survey of Over 200 Medical Technology Companies. November 2010. Retrieved from https://www.advamed.org/sites/default/files/resource/30_10_11_10_2010_Study_CAgenda_makowerreportfinal.pdf

[7]Makower J., Meer A., Denend L., FDA Impact on U.S. Medical Technology Innovation: A Survey of Over 200 Medical Technology Companies. November 2010. Retrieved from https://www.advamed.org/sites/default/files/resource/30_10_11_10_2010_Study_CAgenda_makowerreportfinal.pdf

5.1.2 PMA Product

For higher-risk devices seeking premarket approvals (on the PMA pathway), responding companies indicated that it took an average of 54 months to work with the FDA from first communication to being approved for marketing the device. In Europe, it took an average of 11 months from first communication to approval.

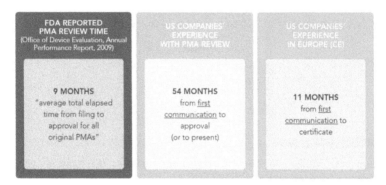

Figure 5.3 PMA and CE Mark regulatory timelines. Reproduced with permission from footnote 8.

It took approximately 32 months to develop a product and around 1.5 years to develop a clinical unit (see Fig. 5.4). The clinical and regulatory phases took about 6 years.

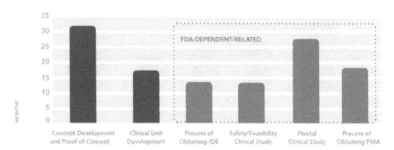

Figure 5.4 Average time by the stage for PMA product. Reproduced with permission from footnote 9.

[8]Makower J., Meer A., Denend L., FDA Impact on U.S. Medical Technology Innovation: A Survey of Over 200 Medical Technology Companies. November 2010. Retrieved from https://www.advamed.org/sites/default/files/resource/30_10_11_10_2010_Study_CAgenda_makowerreportfinal.pdf

[9]Makower J., Meer A., Denend L., FDA Impact on U.S. Medical Technology Innovation: A Survey of Over 200 Medical Technology Companies. November 2010. Retrieved from https://www.advamed.org/sites/default/files/resource/30_10_11_10_2010_Study_CAgenda_makowerreportfinal.pdf

Important Takeaways:
- Bringing the device to market requires multiple years.
- Clinical and regulatory phases can be twice as long as product development itself.

5.2 How Much Money Do You Need?

5.2.1 Class II Devices

The survey data found that the average total cost for participants to bring a 510(k) product from concept to clearance was approximately $31 million, with $24 million spent on FDA-dependent and/or related activities. The product development costs were $4 million, and another $3 million were associated with clinical unit development. These numbers are summarized in Fig. 5.5.

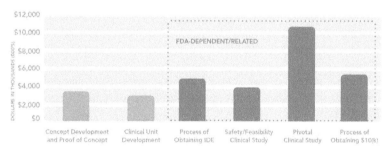

Figure 5.5 Average total expenditure per stage for 510(k) products. Reproduced with permission from footnote 10.

These results are supported by findings from Chin et al. for microfluidic point-of-care IVDs.[11] Based on their numbers it can be estimated that six years and $34 million are required on average to deliver a point-of-care device. With one exception, most companies have raised between 10 and 50 million US dollars by the time their first POC diagnostics product has been approved

---

[10]Makower J., Meer A., Denend L., FDA Impact on U.S. Medical Technology Innovation: A Survey of Over 200 Medical Technology Companies. November 2010. Retrieved from https://www.advamed.org/sites/default/files/resource/30_10_11_10_2010_Study_CAgenda_makowerreportfinal.pdf

[11]Chin C. D., Linder V., Sia S. K. (2012), Commercialization of microfluidic point-of-care diagnostic devices, *Lab on a Chip*, **12**, 2118–2134.

5.2.2 Class III Devices

For PMA devices, the survey found that the average total cost from concept to approval was $94 million, with $75 million spent on stages linked to the FDA. The product development costs were around $10 million, and another $9 million were associated with clinical unit development. The results are depicted in Fig. 5.6.

Note that the estimates from the survey do not include the cost of obtaining reimbursement or any sales/marketing-related activities.

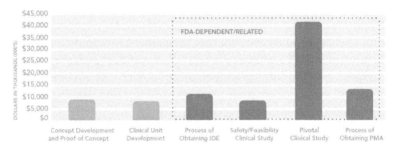

Figure 5.6 Average total expenditure per stage for PMA product. Reproduced with permission from footnote 12.

These numbers are supported by Rao,[13] who compiled data for companies in implanted devices in the ophthalmological space (see Table 5.1)

Rao[14] also offered a rule of thumb that the total funding is 3× of the clinical studies' cost. He split the product development into three phases: Innovation, Safety, and Regulatory and assign $5 million, $40 million, and $50 million to each phase. Thus, around $95 million are required to bring to the market PMA device in the ophthalmological space.

---

[12]Makower J., Meer A., Denend L., FDA Impact on U.S. Medical Technology Innovation: A Survey of Over 200 Medical Technology Companies. November 2010. Retrieved from https://www.advamed.org/sites/default/files/resource/30_10_11_10_2010_Study_CAgenda_makowerreportfinal.pdf

[13]Rao R. LensGen, The True Cost of Developing a Medical Device, in Medical Device Playbook, StarFish, Oct 27, 2020.

[14]Rao R. LensGen, The True Cost of Developing a Medical Device, in Medical Device Playbook, StarFish, Oct 27, 2020.

Table 5.1 Funding raised for ophthalmological devices. *Source*: Adapted with permission from footnote 15

Company	Total Capital	Status and Comments
PowerVision	$130M	Pilot Studies only. Not close to IDE Submission
RxSight	$140M	US clinical studies; raised additional $75M for approval
Acufocus	$235M	IC-8 IOL, in US clinical studies
ReVision	$195M	FDA approval; partial commercialization
Ivantis	$132M	FDA approval (glaucoma stent)
Transcend	$98M	FDA approval (glaucoma stent)
Glaukos	$121M	FDA approval; incubation capital is unknown
VisioGen	$86M	Completed FDA pivotal trial; Study aborted
LensGen	$42M	Plan: $92M through approval of two products

5.3   Exits

Data from public sources also provide some insights about potential exits. As discussed in Chapter 1, two primary ways to cash out the startup are through either initial public offering (IPO) on a stock exchange or merger and acquisition (M&A) with another market participant.

Each of these primary pathways has multiple variations. For example, so-called reverse takeovers (or reverse mergers) particularly through special purpose acquisition companies (SPAC) gained popularity recently. While IPOs typically happen for companies with revenue and product on the market, M&As may occur at the pre-commercialization stage. For example, my company was acquired at the pre-regulatory phase.

As we mentioned in Chapter 1, before 2010, statistics in MedDev space was skewed toward M&As: more than 90% of corporate transactions in the MedDev space were M&As. However, most current data shows that this ratio is close to 2:1 now (see Table 1.1).

---

[15]Rao R. LensGen, The True Cost of Developing a Medical Device, in Medical Device Playbook, StarFish, Oct 27, 2020.

5.3.1 IPOs

In Table 5.2, one can see the statistics for recent exits through IPOs. There is a notable increase in pre-money IPO valuations. This increase is in line with raising evaluations of investment rounds, which we will discuss in Chapter 6.

Table 5.2 IPO statistics in MedDev space: 2014-2020. *Data source*: Footnote 16

Year	Number of IPOs	Median Pre-money	Median Raised
2014	10	$189M	$70M
2015	11	$163M	$72M
2016	3	$164M	$75M
2017	3	$59M	$26M
2018	8	$217M	$86M
2019	8	$297M	$94M
2020	11	$469M	$149M

5.3.2 M&As

In Table 5.3, one can see the statistics for recent exits through M&As. Note a significant increase in milestone payments (the difference between the total deal and upfront payment).

Table 5.3 M&A statistics in MedDev space: 2014–2020. *Data source*: Footnote 17

Year	Number of M&As	Upfront Payment	Total Deal	Years to Exit
2014	18	$180M	$185M	6.9
2015	19	$125M	$141M	7.0
2016	12	$173M	$260M	8.6
2017	14	$131M	$283M	7.7

[16]Silicon Valley Bank (2021), Healthcare Investments & Exits Annual 2021 Report. Retrieved from https://www.svb.com/globalassets/library/managedassets/pdfs/healthcare-report-2021-annual.pdf

[17]Silicon Valley Bank (2021), Healthcare Investments & Exits Annual 2021 Report. Retrieved from https://www.svb.com/globalassets/library/managedassets/pdfs/healthcare-report-2021-annual.pdf

Year	Number of M&As	Upfront Payment	Total Deal	Years to Exit
2018	20	$195M	$223M	7.8
2019	17	$120M	$220M	7.9
2020	16	$132M	$210M	4.8

5.3.3 Regulatory Stage at Exit

Table 5.4 slices M&A data from another perspective. It provides a view on regulatory stage of the company during exits (Private M&As). In particular, it should be noted that most of the deals in Class III devices are pre-FDA approval. Thus, the costliest regulatory approval phase ($50 million according to Rao[18]) has not occurred yet.

Table 5.4 Number of M&A deals by a pathway. *Data source*: Footnote 19

Year	Total	CE Mark	FDA	Pre-Commercial
2015	$3.7B	5	10	4
2016	$4.0B	4	7	1
2017	$4.7B	3	8	3
2018	$5.1B	3	15	2
2019	$9.2B	2	13	2
2020	$3.8B	2	9	5

5.3.4 Probability of a Successful Exit

Based on the market statistics, we can assess the probability of a large successful exit in the MedDev space.

On the exit side, we have approximately 25 large exits a year (see Table 5.2 and Table 5.3). Assuming the steady pipeline of companies, we can calculate the average number of companies in the pipeline.

---

[18]Rao R. LensGen, The True Cost of Developing a Medical Device, in Medical Device Playbook, StarFish, Oct 27, 2020.

[19]Silicon Valley Bank (2021), Healthcare Investments & Exits Annual 2021 Report. Retrieved from https://www.svb.com/globalassets/library/managedassets/pdfs/healthcare-report-2021-annual.pdf

This number can be calculated in several ways. The survey[20] assessed the number of MedDev companies as 1023, while FDA figures show 4776. Most likely, these numbers can be interpreted as the following. The first number is the number of companies, which received institutional investments. The second number is the total number of companies in place, including underfunded ones. Assuming six years in the clinical and regulatory phases (Figs. 5.2 and 5.4), we can conclude that the probability of exit for the funded company is around 15%. For the entire pool of companies (including underfunded), the probability can be assessed as 3%.

The chances for a successful exit increase as the company goes further through the investment rounds. The number of late-stage companies can be estimated from the number of deals. As we will discuss in Chapter 6, there are approximately 300 MedDev deals a year, with about 1/3 of them are late-stage. Assuming 18 months between the funding rounds, we can conclude that there are around 150 active late-stage MedDev companies. Taking five years as an average age for the late stages, we can expect that the probability of successful exit for a late-stage MedDev company is around 83%.

5.4 Conclusions

The MedDev product development is a lengthy and costly process. On average, it takes more than eight years to bring the device to the market, and it costs from $31 million (Class II device) to $95 million (Class III device).

While these numbers are based on 2010 data (study[21]), they are confirmed by several other sources.

In Table 5.5, one can see the funding amount raised and years to approval for a mix of Class II and Class III devices compiled by Drlik[22] from public sources (CrunchBase).

---

[20]Makower J., Meer A., Denend L., FDA Impact on U.S. Medical Technology Innovation: A Survey of Over 200 Medical Technology Companies. November 2010. Retrieved from https://www.advamed.org/sites/default/files/resource/30_10_11_10_2010_Study_CAgenda_makowerreportfinal.pdf

[21]Makower J., Meer A., Denend L., FDA Impact on U.S. Medical Technology Innovation: A Survey of Over 200 Medical Technology Companies. November 2010. Retrieved from https://www.advamed.org/sites/default/files/resource/30_10_11_10_2010_Study_CAgenda_makowerreportfinal.pdf

[22]Drlik M., Starfish Medical, "How Much Does it Cost to Develop a Medical Device?" Retrieved from https://starfishmedical.com/assets/StarFish-Whitepaper-Cost-to-Develop-Medical-Devices-July-2020.pdf

Table 5.5 Funding raised and years to approval for devices. *Source:* Reproduced with permission from footnote 23, Copyright StarFish Product Engineering

Device	Design 'complete'	Human clinical studies required/ completed	Regulatory clearance and classification	Commercial sales	Time from initial to final fundraise during development[3] (years)	Amount raised
Ablation device	Yes	Yes/Yes	Yes-FDA Class III	Yes	9.5	$77 M
Vision correction device	Clinical prototype	Yes/No	Yes-FDA Class III	No	4.5	$ 29.6 M
Eye implant	Yes	Yes/In Progress	Yes-FDA Class III	No	6	$ 42 M
Optical cart-based system	Yes	Yes/Yes	Yes-FDA Class II	Pending	N/A	$ 5 M
Software as a medical device	Yes	No	Yes-FDA Class II	Yes	N/A	$ 3.2 M
Diagnostic device	Yes	No	Yes-N/A	Yes	5	$ 24.6 M
Blood treatment device	Yes	Yes/Yes	Yes-FDA Class II	Pending	N/A	$ 500 M[4]

[23]Drlik M., Starfish Medical, "How Much Does it Cost to Develop a Medical Device?" Retrieved from https://starfishmedical.com/assets/StarFish-Whitepaper-Cost-to-Develop-Medical-Devices-July-2020.pdf

(Continued)

Table 5.5 (*Continued*)

Device	Design 'complete'	Human clinical studies required/ completed	Regulatory clearance and classification	Commercial sales	Time from initial to final fundraise during development[3] (years)	Amount raised
Point of care (PoC) blood assessment	Yes	No	Yes-FDA Class II	Yes	6.5	$ 17.3 M
Light therapy	Yes	No	Yes-FDA Class II	Yes	N/A	$ 14 M
Cell processing	Yes	?	Yes-N/A	Yes	N/A	$ 15 M
Novel ultrasound device	Yes	No	Yes	Yes	5	$ 20.5 M
Point of care (PoC) microfluidic assay system	Yes	No	?	Yes	9	$ 84 M
Average					**6 Years**	**$ 25.5 M**

As we mentioned before, not all 510(k) devices require clinical validation. In this case, the regulatory costs (Fig. 5.5) drop dramatically, and the device can be developed and obtain regulatory clearance for a fraction of cost: $4–5 million.

PART II
IMPORTANT INGREDIENTS

In Part I, we talked about the product development roadmap and its most important stages. We found that product development in the MedTech world is very different from any other market sector. These differences have major implications on all aspects of the startup lifecycle.

We discuss these factors in this part of the book. The focus of Part II is to delineate primary points on how to implement the product development roadmap.

We will start with funding. Funding in the MedTech startup world is very different from any other startup. While you can brainstorm an idea of a great consumer app with friends, develop the rapid prototype over the weekend, and start pitching it to investors a week later, this approach does not work in MedTech, and particularly for medical devices. It takes years to eliminate technology and regulatory risks and make investors comfortable. Thus, investor funding is a relatively remote opportunity for virtually any MedTech venture. Instead, the MedTech ventures rely initially on grant funding. In Chapter 6 (Funding), we will discuss grant and equity funding for MedTech startups.

Any idea worth replication will be replicated. Thus, to survive and become successful, the business needs to find a competitive advantage and protect it. Competitive advantages and their protection will be the focus of Chapter 7 (IP and Other Moats).

Not many scientists have a good understanding of business and entrepreneurship. Fortunately, in recent years some tools appeared to simplify and streamline entrepreneurship. Thus, we start Chapter 8 (Business Model) with a brief explanation of the Lean Canvas.

Healthcare has a notoriously complex structure. So, it is quite challenging to develop a business model successfully integrated into this complex system. To do it, the MedTech entrepreneur needs a reasonably good understanding of how various pieces of this system fit together and the motivation for each player. It will be discussed briefly in the second part of Chapter 8.

Financing and business model are essential ingredients of the MedTech startup success. However, they are not the only important factors. Research shows at least three other factors (team, quality of the idea, and timing) are even more critical. These and some other factors will be considered in Chapter 9 (Other Considerations).

We finish the main body of the book with Chapter 10 (Silver Lining). The whole book, I tried to explain why MedTech is difficult. However, it is not all doom and gloom. In Chapter 10, we briefly consider why MedTech is a great place to be and why we, scientists, are well-positioned to be there.

Chapter 6

Funding

A medical device startup is a long journey. Thus, the availability of money is a necessary condition for survival and thriving. Actually, one of the primary job functions and skills of any startup CEO is finding the money.

6.1   Types of Funding

There are multiple types of funding available for startups in general and MedDev startups in particular. Different types of funding are typically available at various stages of the startup lifecycle. They also come from multiple sources: grant agencies, angel investors, VCs, etc. Thus, the conditions, amounts, and strings attached to each type are entirely different.

The startup can be funded through three major types of funding: grants, debt, and equity. We will consider them in some detail, with examples primarily coming from North America (the United States and Canada).

Grants are a non-dilutive form of funding. Most often, grants are provided by government agencies, but multiple non-profit and for-profit organizations also offer grants. Typically, this type of funding has the smallest amount of strings attached.

Bringing a Medical Device to the Market: A Scientist's Perspective
Gennadi Saiko
Copyright © 2022 Jenny Stanford Publishing Pte. Ltd.
ISBN 978-981-4968-25-6 (Hardcover), 978-1-003-31221-5 (eBook)
www.jennystanford.com

Equity is a primary way of funding at later stages of a startup. In an equity round of funding, an investor (or most often a group of investors) provides funding in exchange for shares (ownership) in the company.

Two major types of participation in the company are though common and preferred shares.

When people talk about stocks, they usually refer to common stocks, representing shares of ownership in a corporation, including voting rights.

Preferred shares, in general, do not have voting rights. They also have a fixed rate of return, which makes them similar to debt. However, preferred shares come in many flavors. For example, convertible preferred shares can be converted into common stock at a fixed conversion ratio.

However, the primary difference between common and preferred shareholders is the order in which they received funds upon their availability, e.g., during liquidation of the company. The common shareholders are the last in line, while the preferred shareholders are ahead of common shareholders but behind the debt holders.

That feature of preferred shares is the primary reason for their popularity in the startup environment. They allow putting investors ahead of founders, employees, advisors, etc., which have common shares. Preferred shares in startups typically have multiple features attached (including voting), making them quite similar to common shares in all other aspects other than the priority of distribution.

This brings us to the last type of funding: debt. Technically, debt is the least feasible and suitable source of funding for startups. Startups are very risky ventures with a very low probability of success and return of money. It is precisely the opposite of what lenders are looking for in companies. Moreover, startups (particularly life sciences and MedDev) can be pre-revenue for many years. Thus, it is almost undoubtedly long-term financing. That is why it is extremely difficult for a startup to receive a loan from banks, which rely on cash flow to assess an ability to serve the debt. If granted, the loan will require collateral such as personal guarantees.

However, in some cases, the short-term funding is still feasible. It is mainly possible when it serves as a bridge to finance some activities (e.g., clinical trial) in anticipation of the next round of funding. Such funding can be provided by venture debt providers (e.g., Silicon Valley Bank in the United States).

Typically, debt holders are expecting a fixed return. That makes them different for equity holders, who are looking for a significant upside. However, debt holders are the first in line to receive funds upon their availability, e.g., during liquidation of the company. This debt feature created a convertible debt, a hybrid instrument, which became very popular recently. In this case, initially, money is provided to the company as a debt. However, on a particular event (e.g., a round of funding), this debt is being automatically converted into equity using a predetermined formula. In this case, investors have additional protection for some period of time when the company is the most vulnerable. The convertible debt also defers the necessity to agree on the company's valuation, which can be quite challenging at the company's early stages.

A similar but simpler alternative to convertible debt in a startup is a SAFE (simple agreement for future equity), popularized by Y Combinator. A SAFE is an agreement between an investor and a company that gives rights to the investor for future equity in the company. It is similar to an option or warrant, except without determining a specific price per share at the time of the initial investment.

As convertible debt and SAFE agreements can be considered mere instruments to simplify investment in pre- and during equity rounds of funding, we will combine this type of funding with equity funding.

We also should note that rounds of funding are often accompanied by issuing warrants or options, which are used to sweeten the deal. They do not bring immediate economic value and can create a substantial dilution in the future.

6.2   Sources of Funding

Equity funding other than the friends and family type is typically not available for very early-stage startups. That is why for many MedDev startups, the first funding coming from grants.

6.2.1 Grants

If you search "grant" on the Government of Canada website, you will find several hundred grant opportunities. Nowadays, governments provide multiple options for grant funding. Their primary interest is job creation, and they see startups as an instrument in it. To emphasize the importance of startups for the economy, South Korea even has the Ministry of SMEs and startups.

Government funding comes in different flavors. Governments have programs for specific industries, specific territories, and particular demographics. In addition to government agencies, there are multiple grant opportunities from non-government sources.

Typically grant programs are tailored to a specific phase of the product development.

6.2.1.1 Research grants

Research grants are probably the best way to fund a MedTech venture at the very early stages. In this case, the company's primary goal is to develop technology, which aligns with research grants.

One of the best parts of research grants is that they are typically pretty significant in size. However, they usually fund only some aspects of the product development, and eligible expenses are quite limited.

They also differ in the type of research they support. Some grants focus on basic science, others on clinical science or translational research. Here, there are several examples of Health Science research grants from the United States and Canada.

In the United States, the primary granting agency for Health Sciences is the National Institutes of Health (NIH). There are 21 separate institutes and 6 centers in this system. They provide funding for researchers and small businesses. The primary grants that can be used to fund technology development are:

- R01- NIH Research Project Grant Program
  - R01 supports a discrete, specified, circumscribed research project and is generally awarded for 3–5 years. No specific dollar limit unless specified in a funding opportunity announcement (FOA). However, advance permission is required for $500K or more (direct costs) in any year.

- R21 - NIH Exploratory/Developmental Research Grant Award
  - R21 is used for new, exploratory, and developmental research projects and sometimes used for pilot and feasibility studies. It is limited to up to two years. The combined budget for direct costs for the two-year project period usually may not exceed $275,000.

A primary source of research funding for post-secondary institutions across Canada is Tri-Agency: The Natural Sciences and Engineering Research Council of Canada (NSERC), the Canadian Institutes of Health Research (CIHR), and the Social Sciences and Humanities Research Council (SSHRC). The health research is funded primarily by CIHR and NSERC.

These agencies focus on different aspects of the research. NSERC focuses on scientific and engineering areas. CIHR focuses on four health research areas: biomedical, clinical, health systems services, and population health. Thus, for example, NSERC may fund basic science and engineering aspects of the medical device. However, it will not fund clinical studies of such devices, which can be financed through CIHR grants.

The Project Grant program is the primary research funding available from CIHR. It supports projects or teams/programs of the research proposed and conducted by individual researchers or groups of researchers in all health areas.

In addition to pure agency funding, there are cross-agency projects. Collaborative Health Research Projects (CHRP) is a joint initiative between the NSERC and the CIHR.CHRP grants support focused, interdisciplinary, collaborative research projects involving any natural sciences or engineering field and any health sciences field.

However, Tri-Agency funding (other than trainee funding) is limited to principal investigators (PIs) only, i.e., university faculty.

6.2.1.2 Industrial research

Several government agencies run Small Business Innovative Research (SBIR) and Small Business Technology Transfer (STTR) programs in the United States. SBIR and STTR are Congressionally-Mandated Set-Aside Programs. Every agency with an extramural research budget greater than $100 Mln per year has to allocate 3.2% of the budget to SBIR. Similarly, every agency with an extramural research budget greater than \$1 Bln per year has to allocate 0.45% of the budget to STTR. Several such agencies (NIH, CDC, FDA, DoD, and NSF) have SBIR/STTR programs for human health-related sciences.

NIH is a primary source of industrial research funding in the United States. This funding is available only to small businesses, and NIH administers it through R43/R44—Small Business Innovative Research (SBIR) program.

- R43/R44 is used to support research or research and development (R/R&D) for for-profit institutions. It has a three-phase structure:
  - o I – Feasibility study to establish scientific/technical merit of the proposed R/R&D efforts (generally, six months; \$250,000)
  - o II – Full research or R&D efforts initiated in Phase I (generally two years; \$1,700,000)
  - o IIb – Bridge funding prior to the commercialization stage

There is also an SBIR/STTR program administered by National Science Foundation (NSF). The NSF SBIR/STTR Program funds the development of deep technologies based on discoveries in fundamental science and engineering for profound societal impacts. It has a two-phase structure:

- o Phase I – may not exceed approximately \$250,000 for 6-12 months
- o Phase II – Full research or R&D efforts initiated in Phase I (may not exceed \$1,000,000 over 2 years)

In Canada, industrial research funding is not limited to small businesses only. National Research Council of Canada (NRC) provides the Industrial Research Assistance Program (IRAP). The company needs to provide matching funds (typically 25%) of total eligible expenses.

Note that these grants (particularly for small businesses) are limited to domestic companies only.

6.2.1.3 Academia–industry collaboration

In the United States, NIH is the primary source of academia–industry collaboration funding. They administer it through R41/R42—Small Business Technology Transfer (STTR) program:

- R41/42 are used for cooperative research/research and development (R/R&D) carried out between small business concerns (SBCs) and research institutions (RIs). It has a three-phase structure:
  - I – Feasibility study to establish scientific/technical merit of the proposed R/R&D efforts (generally, one year; $250,000)
  - II – Full R/R&D efforts initiated in Phase I (generally two years; $1,700,000)
  - IIb – Bridge funding prior to the commercialization stage

There is also an SBIR/STTR program administered by National Science Foundation (NSF). The NSF SBIR/STTR Program funds the development of deep technologies based on discoveries in fundamental science and engineering for profound societal impacts. It has a two-phase structure:

- Phase I – may not exceed approximately $250,000 for 6–12 months
- Phase II – Full research or R&D efforts initiated in Phase I (may not exceed $1,000,000 over 2 years)

In Canada, NSERC is the primary source of academia–industry collaboration funding. These funds are administered to academic researchers. There are several programs available to university and college researchers.

- Alliance grant program is designed for university researchers collaborating with private-sector, public-sector, or not-for-profit organizations
  - $20,000 to $1 million per year
  - 1 to 5 years
- Engage grants are intended for college researchers collaborating with private-sector, public-sector, or not-for-profit organizations
  - They are limited to $25,000 for up to 6 months.

In addition to national (federal) programs, there are also multiple regional programs. For example, in Ontario (Canada), the Ontario Centres of Innovation (OCI) administers the Voucher for Innovation Program (VIP), which supports R&D collaborations between companies and publicly funded universities, colleges, and research hospitals. They provide between $20,000 and $150,000 to the post-secondary institution. In this program, the industrial partner needs to provide matching funds (1:1 on a cash basis).

6.2.1.4 Translational grants

In Canada, NSERC administers Idea to Innovation (I2I) Grants. The I2I grants provide funding to college and university faculty members to support research and development projects with recognized technology transfer potential.

6.2.1.5 Commercialization grants

In the United States, NIH provides several grants to commercialize the research.

- R41/42 are used for cooperative research/research and development (R/R&D) carried out between small business concerns (SBCs) and research institutions (RIs). Phase IIb Award provides bridge funding (matched) prior to the commercialization stage.
- R43/R44 is used to support research or research and development (R/R&D) for for-profit institutions. Phase IIb Award provides bridge funding (matched) prior to the commercialization stage.

In Canada, several grant opportunities are available. These funds are typically available to startups on a matching basis.

6.2.1.6 Other grant opportunities

In addition to the standard programs, the government agencies provide multiple ad hoc programs. For example, during COVID-19 pandemic, government agencies offered numerous COVID-related programs.

In addition to the government granting agencies, there are resources available from other government branches, e.g., military and preparedness and response programs.

For example, in the United States, the Biomedical Advanced Research and Development Authority (BARDA), a part of the US Department of Health and Human Services, provides funding for preparedness and response-related projects.

The Defense Advanced Research Projects Agency (DARPA) is a research and development agency of the United States Department of Defense responsible for developing emerging technologies for the military. They provide funding for emergent medical technologies, which can be used in military or veterans' treatment.

Department of Defense also administers several programs, including Peer-Reviewed Medical Research Program (PRMRP). PRMRP addresses multiple diseases relevant to veterans' health.

The US government agencies announce grant funding opportunities through Funding Opportunity Announcements (FOAs). All federal US grants can be found on the Grants.gov site (www.grants.gov). However, projects with high potential can be funded through other proposal intake processes specific to an agency in addition to publicly announced grants.

In addition to government grant agencies, multiple non-government resources are available. They can be split into several buckets:

Research Organization Foundations

Most research hospitals/institutes have their own foundations. They typically provide funding to researchers affiliated with such hospital/institute.

Multinationals

Several big pharma and medical device companies have affiliated incubators and startup support programs. It is a growing trend and becoming a part of virtually any large company's growth plan, especially in High Tech. For example, Johnson & Johnson has Johnson & Johnson Innovation, which runs co-working spaces for MedDev companies (JLABS) in several cities worldwide (e.g., San Diego, New York, Boston, Toronto, and Shanghai). In addition to that, they run regular research challenges on a specific topic.

Patient Groups

A patient group (e.g., American Diabetes Association (ADA)) is an organized group that represents patients with a specific disease or condition, or collection of diseases or conditions, and who come together to help find a cure for it. Depending on the association's size, they provide grants for basic science, translational research, development awards, and trainee awards. Typically these grants are limited to academic institutions and research hospitals only.

Foundations, Philanthropy, and Other Non-profits

There are multiple non-profits, which provide funding to solve healthcare problems. Probably, the best-known global organization is Grand Challenges (grandchallenges.org). It was launched in 2003 by the Bill & Melinda Gates Foundation. Initially, it was established as Grand Challenges in Global Health. Then, it was re-launched in 2014 as Grand Challenges. Its new name reflects its expanded scope encompassing global development challenges; however, healthcare is still its primary vector. They post problem-specific challenges regularly.

Important Takeaways:

- Academia is an ideal place to nurture a MedTech project
- Multiple grant opportunities are available. However, submitting them can be very daunting and time-consuming. Some startups even have a co-founder or other team member dedicated to this job.

6.2.2 Equity Funding

While grants may allow starting developing technology and getting the startup off the ground, equity funding is the primary way to grow and scale startups.

Raising capital is a skill that needs to be mastered by a CEO of any startup, particularly a MedDev startup. This area is covered very extensively, and we will not try to address it holistically. Instead, we refer readers to some good books on the fundraising process and its legal aspects. For life sciences, such resource can be "The Life Science Executive's Fundraising Manifesto" by Dennis Ford.[1]

Here we will discuss the equity funding landscape and its primary players briefly.

---

[1]Ford D., *The Life Science Executive's Fundraising Manifesto*, Life Science Nation, 2014.

6.2.2.1 Friends and family

Friends and family (F&F) are often a primary source of funding during the company's formation and very early stages. Those are people who know you, who want to support you and bet on your abilities.

However, while raising money from friends and family, you need to take into consideration several aspects.

Firstly, you need to be honest with yourself and your F&F about risks and expectations. You will have to deal with these people for many years, irrespective of financial outcomes.

Secondly, an F&F round of funding will not create an "investment history" for your startup. While potential future investors may (and most of them certainly will) appreciate your skin in the game, F&F investment will not significantly increase the probability of the subsequent round of funding.

Aside from F&F, Life Science Nation identified (see footnote 2) 10 types of early-stage life science investors: angel groups, corporate venture capital, foundations and philanthropic, family offices and private wealth, government organizations, hedge funds, alternative institutional investors, large pharma and biotech, private equity, and venture capital. In addition to this list, two more sources can be added: crowdfunding and business incubators and accelerators.

We will briefly cover all these groups in approximate order of their relevance to startup funding (from very early to later stages).

6.2.2.2 Incubators and accelerators

Business incubators and accelerators mushroomed in recent years. They became essential parts of the startup ecosystem.

Business incubators are typically associated with the first step of the startup journey. In many cases, business incubators accept non-incorporated teams of entrepreneurs and help them to launch the company. In some cases, they provide some funding (typically less than $50,000), but in most cases, they offer a collaborative space and other infrastructure only. In most cases, they also don't take any equity in the company.

---

[2]Ford D., *The Life Science Executive's Fundraising Manifesto*, Life Science Nation, 2014.

Most business incubators are campus-based and associated with colleges and universities. Some universities launched multiple thematic business incubators. For example, Ryerson University launched more than 10 business incubators, including the Biomedical Zone. While incubators are typically campus-based, their membership is usually broad and non-limited to a respective university only.

In addition to the working space, incubators typically provide programming and mentorship for new entrepreneurs. Membership in the incubator generally is based on an admission process, where a startup team pitches their idea to a selection committee. Some incubators run time-limited cohorts, where they help to incubate ideas for a limited time, e.g., 4 or 6 months.

Participation in business incubators typically does not have a downside (other than certain challenges maintaining confidentiality in the co-working environment). However, it may have a significant upside. Firstly, it helps first-time founders gain knowledge on various business processes, which some may lack. Secondly, it boosts the credibility of the startup. Such as the idea was selected between others, it gives a perceived rubber stamp of approval for future employees, partners, and investors. Thirdly, it provides some visibility and public relations (PR) presence. Incubators are typically active in the social media space. Finally, in some cases, it gives access to additional funding opportunities. For example, some grant programs are limited to startups associated with certain incubators (e.g., campus-based).

Accelerators target companies, which already got off the ground (typically seed stage). They usually operate using "cohorts"—time-limited pools of startups, typically 3 or 4 months, culminating in a "demo" day, where all cohort's startups present their ideas to outside investors.

If most business incubators are campus-based, the accelerators are typically run by for-profit or not-for-profit organizations. They usually accept the company to the program in exchange for the startup's equity (typically 6%). Some of them provide some nominal funding, which is included in the mentioned above 6%. Sometimes, funding is available but can be arranged separately in exchange for additional stocks in the company.

Accelerators can be standalone (like Y Combinator). However, some companies run multiple accelerators, often across various geographical locations. Examples of such chains can be Startupbootcamp or Techstars. In this case, each individual accelerator is typically theme-based, e.g., digital health or FinTech.

Participation in accelerators may have its upsides and downsides. Upsides are numerous. Firstly, it may give access to partners and industry experts. Secondly, it further boosts the credibility of the startup. Such as the idea was selected between others, it provides a perceived rubber stamp of approval for future employees, partners, and investors. The caliber of the accelerator is crucial here. Participation in a renowned program like Y Combinator will certainly boost credibility. Thirdly, it gives some visibility and public relations (PR) presence. Accelerators are active in the social media space. Finally, it may provide access to additional funding opportunities. For example, a "demo" day can be an excellent opportunity to present an idea to multiple investors.

The most significant downside of accelerators is that the startup may receive very little of the mentioned above benefits in exchange for dilution of its ownership structure. Not all accelerators are created equal. Research and vetoing are very important.

The common theme of both incubators and accelerators is working in a collaborative space (co-working). It brings several other benefits. Firstly, it is access to a pool of people and ideas. Secondly, it is an immersion in an atmosphere of motivated and like-minded people. Some companies, even when they start generating revenue, prefer to stay in such an environment.

Key Takeaways:

- Business incubators typically work with un-incorporated or recently incorporated startups. They provide programming and mentorship and usually do not take any equity stake in the company.
- Accelerators typically work with seed-stage companies. They provide programming and mentorship in exchange for the equity stake.

6.2.2.3 Crowdfunding

Crowdfunding emerged recently as an alternative source of funding.

There are two distinct models of crowdfunding. The first model, which is more commonly known, consists of donations in exchange for early access to the product, discounts, or tokens of support. The most notable platforms, which use this mechanism, are Indiegogo and Kickstarter.

The second mechanism is equity funding. In this case, each contribution is an investment in exchange for a share in the company. This mechanism was introduced in the United Staes in 2011 by Jumpstart our Business Startups (JOBS) Act.

There are several common themes between crowdfunding platforms.

Firstly, they charge significant fees (in the 5%–20% range) of the total amount raised. Secondly, successful fundraising is quite a challenging process in general. To get any funding, the project requires visibility. However, to get visibility on the crowdfunding platform, the project needs to get promoted by the platform's algo. This means hitting certain time-sensitive performance targets, which can be quite challenging on its own.

From a historical perspective, most MedTech projects do not perform well on the general crowdfunding platforms. In order to raise money successfully, the project needs hype and gain quick momentum. However, most "crowd" members (backers) do not understand MedTech projects' intricacies. Moreover, the timeline of the delivery of medical devices is quite long, which is not aligned with the expectations of many members of the "crowd." So, virality is hard to achieve. Times of public emergency are the notable exemption here. For example, during COVID-19 pandemic, there were multiple successful personal protective equipment projects.

To address specific challenges of the crowdfunding for life sciences and MedTech companies, certain dedicated healthcare platforms, e.g., Medstartr, emerged. Medstartr evolved its business model since it started in 2008. While initially, they began as a donation-based platform, now they have an investment-based component (medstartr.vc).

While donation crowdfunding represents certain opportunities for MedTech founders, it also poses significant challenges from a legal perspective. Namely, medical devices are regulated products. Thus, their marketing is allowed upon regulatory clearing or approval. Consequently, all marketing materials (including descriptions on the crowdfunding platform) need to be carefully crafted.

Overall, it seems that donation-based crowdfunding applications are limited to healthcare IT, well-being projects, and MedTech projects, which deliver products for patient use. This strategy can be used to demonstrate user traction at some infliction points. However, most likely, it cannot serve as a long-term strategy.

While most crowdfunding happens through a donation-based process, some opportunities in equity-based (or investment-based) crowdfunding exist. Even though equity crowdfunding represents certain opportunities for founders, it also poses significant challenges, which stem primarily from the fact that equity-based crowdfunding is a regulated process. For example, it is regulated by the Securities and Exchange Commission (SEC) in the United States. Thus, equity-based crowdfunding in the US can be done only through equity portals registered with SEC.

Firstly, startups in equity-based crowdfunding are subject to significant regulatory scrutiny. For example, a crowdfunding platform needs to perform due diligence on the startup and background check on its officers. The financial statements need to be prepared ($100,000–$500,000 raised) or audited (>$500,000 raised) by an independent certified public accountant, which can add to administrative costs significantly. Secondly, the startups in equity-based crowdfunding are subject to ongoing disclosure requirements. With a large number of investors, it can be challenging to maintain confidentiality. Thirdly, the maximum amount, which can be raised, is limited. For example, under the JOBS Act, the aggregate amount of interests sold through an investment-based platform in any 12-month period may not exceed $1 million. Finally, if a company successfully raised funds through investment-based crowdfunding, it may pose significant investor relations problems in the future. A large number of small investors, typical for equity-based crowdfunding, may create serious practical and administrative issues while raising the next funding rounds.

Key Takeaways:

- There are two models of crowdfunding: donation-based and equity-based.
- Donation-based crowdfunding can be challenging for most MedTech projects.
- Equity-based crowdfunding is an investment process regulated by financial market regulators. Even if the round was successful, it might pose significant ongoing challenges for the early-stage startups.

6.2.2.4 Angel investors

In many cases, angel investors are the first "professional" investors for startups. Angel groups (or syndicates) are the groups of high-net-worth individuals interested in direct investments into early-stage companies.

The typical size of angel group investment is from $200,000 to $2,000,000. This amount is split between several group members, which decreases the overall risk for each member. Angel groups also leverage their infrastructure and collective expertise for due diligence.

Angel groups have become very popular nowadays. According to footnote 3, in 2014, around 100 active groups were investing in BioTech and MedDev globally. As of the time of writing, there were more than five groups in Southern Ontario alone. However, dealing with angel groups, you need to aware of certain aspects associated with them.

Firstly, typically angel groups have a very narrow geographical focus. They invest only in their own backyard.

Secondly, angel groups are loose groups of individuals. It may make the negotiation process and investor relationships quite complicated. In addition to that, the overall investment process can be very long (more than a year can be quite typical).

Finally, typically angel groups are not industry-focused. It means two things: (a) they may lack interest/expertise in the MedDev industry, and (b) even if they are interested, your idea competes with hundreds of other ideas from non-regulated sectors, which can be much more financially appealing. Notable exceptions here

---

[3]Ford D., *The Life Science Executive's Fundraising Manifesto*, Life Science Nation, 2014.

are groups of physicians. Several known angel groups (e.g., Medical Angels in Australia) consist of physicians who invest heavily into MedDev and BioTech early-stage companies.

6.2.2.5 Family offices

The family office refers to a private wealth management vehicle for an ultra-high-net-worth family or individual. Their scope includes wealth management, philanthropy, legal issues, and tangible asset management. Many family offices have a net worth of more than $125 Mln (which is the definition of an ultra-high-net-worth individual), and some of them more than $1 Bln.

Many family offices have assets set aside to alternative investments and invest into early-stage companies.

There are several aspects, which make family offices particular suitable for BioScience and MedDev startups.

Firstly, family offices are typically engaged in extensive philanthropic activities. The recent trend is to focus less on donations and more on investment with significant social impact potential. BioScience and MedDev startups are perfect targets for such philanthropic investments, especially if the idea could positively impact a global scale.

Secondly, the family may have a history of a particular disease. Thus, they can be highly interested in the development of a breakthrough therapy for such disease.

Finally, some of these families built wealth in particular industries. Thus, they can be interested in identifying strong investment opportunities and be associated with a high caliber scientific breakthrough.

6.2.2.6 Government organizations

In addition to grant funding, government organizations may participate in equity rounds of funding. In most cases, they do not lead the round but rather co-invest.

For example, in Ontario (Canada), the Ontario Centres of Innovation (OCI) has a market readiness co-investment fund. The fund has two streams ($125,000 and $250,000) and co-invests into Ontario-based academic-affiliated companies with disruptive/next-generation technologies that demonstrate early evidence of a scalable, repeatable business model serving an identified need in their market.

6.2.2.7 Foundations, venture philanthropy, and patient groups

While foundations are typically associated with grant funding, most recently, some of them started to provide investments as well. It came with the realization that return from investments may enhance their ability to fund research. Thus, some foundations created investment arms.

Foundations and venture philanthropy groups have overlapping investment interests to drive commercialization and push philanthropic agenda.

An example here can be Toronto Innovation Acceleration Partners (TIAP). TIAP is a member-based not-for-profit organization created by MaRS Discovery District and Toronto-based universities, institutions, and research institutes. Its mandate is to bring the members' most promising research breakthroughs to market.

Patient groups (e.g., American Diabetes Association) are also typically active at grant writing. However, they come with other approaches as well, including investments. In the investment space, traditionally, patient groups partner with foundations and venture philanthropists to co-invest.

However, the current trend is that patient groups take a more active approach and mobilize multiple players to find a cure. What can be of particular interest for life sciences and MedDev companies is that patient groups can use their own networks for clinical studies.

In general, partnering with patient groups has substantial advantages. They are very well connected and typically create their own ecosystem with a focus on the particular disease. For example, they can be particularly helpful during clinical studies.

6.2.2.8 VCs

Venture capital firms are the backbone of the equity funding for virtually any startup, including MedTech space. VCs provide an unproportionally large share of total funding. These are large, sophisticated investors, which have an excellent understanding of the marketplace.

However, it should be noted that they typically invest in post-seed rounds of funding: A, B, C, etc.

6.2.2.9 Corporate venture capital

Corporate venture capital is a strategy that multiple large corporations have adapted to allow them to have a hand in emerging technologies that are either directly or indirectly aligned with their primary business. Thus, in a nutshell, corporate venture capital is venture capital funds backed by large corporations.

This type of investor can be of particular interest to BioTech and MedTech startups because they may access large corporations' resources and expertise. For example, Microsoft (which invests in HealthTech) has an extensive partnering program, including access to their extensive sales network (co-sell).

6.2.2.10 Big Pharma and biotech

The paradigm shift is also happening in big pharma and biotech R&D space. Traditionally, these firms had huge budgets for in-house R&D. Nowadays, it is getting hard to compete for top talents, many of which opt to work in a startup environment or start their own companies.

In response, big companies reduced the in-house R&D budgets and filled the gap in their pipelines by acquiring BioTech startups to adapt to these changes. Thus, instead of acquiring later-stage companies (the past strategy), big pharma extended its scope to earlier-stage companies (Clinical Phase I and II).

6.2.2.11 Pension plans and endowments

Pension plans and endowments are also recent newcomers to the startup space. They are highly sophisticated institutional investors with huge assets under management. Their long-time horizon is another advantage. It is aligned with VCs and is suitable for MedTech companies. Traditionally, they have strict investment mandates. However, recently, they started experimenting with alternative investments, including direct investments in life sciences. Given their overall tremendous assets under management (AUM), even relatively small allocations percentage-wise are still massive money-wise.

Pension plans typically invest in later-stage companies.

Endowments typically invest in companies, which stemmed from the affiliated institution. However, occasionally, they can invest in non-affiliated startups. Endowments have a long-term investment horizon so that they may be more comfortable in investment in companies at earlier stages.

6.2.2.12 Hedge funds

Hedge funds are relatively new to the startup environment. Historically, they invest in public markets. However, under the increased pressure from investors to boost their performance, some hedge funds are looking into private equity opportunities and may be open to investing in startups. However, primarily they might be interested in the later stage (e.g., post-revenue) startups and cannot be considered for very early-stage funding.

6.2.2.13 Private equity

The traditional focus of the private equity (PE) firms is established post-revenue companies. PE funds restructure them and optimize their performance. However, similar to hedge funds, PEs quickly adapt their strategies to a booming startup environment. The primary approach that they employ is to aggregate several smaller companies, grow, and optimize the portfolio. The portfolio can be resold at a later stage to a strategic investor to fill the pipeline gap. PE firms leverage their core competency in restructuring, optimization, and operational efficiency in implementing this strategy. Thus, while working with PE firms, founders need to be sure that their visions are aligned with investors' ones.

Note that some investors that we mentioned in the current discussions are buyers rather than investors. For example, in many cases, private equity funds and big pharma will be more interested in acquiring the company than in investment in exchange for non-controlling stock of shares.

6.2.3 Other Sources

There are multiple other sources of funding for the startups. We will briefly mention several of them.

6.2.3.1   Wage subsidies

Government agencies offer various wage subsidies. These subsidies are typically tailored to support the training or hiring of the youth.

For example, in Canada, the NRC-IRAP runs the Youth Employment Program, which "offers financial assistance to offset the cost of hiring young talent to work on projects with R&D, engineering, multimedia, or market analysis components or to help develop a new product or process."[4]

6.2.3.2   Tax incentives

Various branches of government offer multiple tax incentives to support particular activities or industries.

For example, in Canada, the Scientific Research and Experimental Development (SR&ED) Program gives tax incentives to encourage Canadian businesses to conduct domestic research and development (R&D). According to the Canada Revenue Agency (CRA), which runs this program, these tax incentives come in three forms: an income tax deduction, an investment tax credit (ITC), and, in certain circumstances, a refund. The SR&ED Program provides more than $3 billion in tax incentives to over 20,000 claimants annually.[5]

A similar program exists in the United Kingdom. Research and development (R&D) reliefs support companies that work on innovative projects in science and technology.

6.2.3.3   In-kind contributions

Some in-kind contributions can be of great value for startups. For example, incubators, accelerators, and other providers offer mentorship programs for entrepreneurs. It can be of great importance for first-time founders. While in many cases, they typically provide some general business mentorship, in some cases, it can be from executives with a life sciences background, which can have an even higher value for MedTech entrepreneurs.

---

[4]NRC, NRC IRAP funding to hire young graduates. Retrieved from https://nrc.canada. ca/en/support-technology-innovation/nrc-irap-funding-hire-young-graduates
[5]CRA, Scientific Research and Experimental Development Tax Incentive Program. Retrieved from https://www.canada.ca/en/revenue-agency/services/scientific-research-experimental-development-tax-incentive-program/overview.html

The other example of in-kind contribution can be access to a patient network from a patient advocacy group. It may save hundreds of thousands or even millions of dollars in clinical trial expenses.

6.3 Funding and Startup Lifecycle

As one can see from the previous discussion, different types of players write grants or invest at various startup lifecycles stages. This information can be aggregated in a single chart for easy visualization. This type of investor landscape chart is better known for life sciences startups so that we will reproduce here a traditional "life science" investor landscape (see Fig. 6.1).

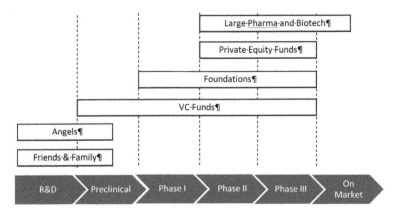

Figure 6.1 Funding sources along with product development life cycle in life sciences.

As you may notice from the chart, there is a significant gap in fund availability during the pre-clinical stage. It is known as a "death valley." F&F and angels' money can be insufficient to complete the pre-clinical stage; however, the venture is still not on a radar screen for most VC funds.

There are several primary drivers for such disconnect.

Firstly, it is connected with risk tolerance. R&D and pre-clinical stage startups are associated with considerable technology risks. Thus, the probability of failure is very significant. Most investors are comfortable with commercialization risks, but they look for opportunities where other risks (particularly technology risks) are eliminated or substantially reduced.

Secondly, it is associated with the concept of assets under management. It can be elaborated on a hypothetical VC fund example. For example, if the fund has $1Bln under management, it can invest $1Mln in each of 1000 companies, $10Mln in each of 100 companies, or $100Mln in each of 10 companies. While splitting investments between more companies decreases individual risks, it also increases the administrative complexity and decreases leverage that the fund may have in each company (e.g., participating in the board of directors). Thus, many funds limit their investments to 20–30 companies. In this case, they can track investment and exercise control more closely. Consequently, an average investment for such a fund will be in the $30Mln–$50Mln range, limiting the type of deals they are looking for ($100Mln and more in pre-money valuation).

Equity investments look like a better investment alternative than debt with low interest rates (which takes place for 10+ years already as of the time of writing). Thus, money migrated from debt into equity asset pools. VC funds represent a tiny fraction of the total equity investment universe, leading to their swelling. According to Crunchbase, in 8 years (2012–2019), venture funds received an astonishing $1.5Trln globally. It brought some VCs to the "Money as a Moat" philosophy when investors flush startups with money to achieve competitive advantage. Thus, we have two types of investors: individual investors (like angel groups) with assets on the scale of millions of dollars and institutional investors (e.g., some VC funds, pensions, and endowments), with assets on the scale of billions of dollars.

As a consequence, we have a paradox. Startup investors have a lot of money available. However, these resources are available for later-stage ventures only.

Some industry experts argue that this landscape has been changed, and multiple other players cover almost the whole product development lifecycle. The reason often cited for that is the Great Recession of 2008–2009, which eliminated numerous VC funds. Thus, other players (family offices, hedge funds, venture philanthropy, and patient groups) came to fill the gap. Perhaps, it was true for several years immediately after the 2008–2009 crisis. However, now it is not the case anymore. VCs are at the helm of the game again, and other players (hedge funds, pensions, endowments) may have even more significant amounts of assets under management.

Thus, the bulk of MedTech startups have very really serious challenges with funding. At the early stages, risks are high, and the investment horizon is very long. Moreover, their output will probably not be as stellar as a life sciences startup with a miracle drug. According to footnote 6, it is only one year in 2000–2019 when a device-focused VC strategy outperformed the biotech/pharma strategy. The average differential over the period has been 22.7% in favor of biotech/pharma. Thus, private investors have very little appetite for MedDev.

A natural solution to this problem could be an extension of government funding. Some countries, like the European Union and Singapore, are quite active in this direction. But North America, with its strong belief in a "magic hand" of a free market, is lagging in it.

To quantify our discussion, let's finish this part with investment statistics.

In Table 6.1, one can see medical devices in the context of VC healthcare funding for the United States and Europe. Several important observations can be deduced from Table 6.1. Firstly, HealthTech, diagnostics, and devices account for approximately 20%, 18%, and 12% of overall VC funding for healthcare ventures. Secondly, the overall financing of these sectors exceeds $20B. Thirdly, the United States accounts for more than 80% of the funding in the MedTech space.

In Table 6.2, one can see the different slices of the MedTech deal flow: number of deals and average deal size.

However, the average investment deal size is not very informative. It includes the whole spectrum of deals from seed to late-stage ones. In order to get a more granular picture, we can extract several insights from other market statistics:[7]

- The number of deals is roughly split equally between first-time, early-stage, and late-stage for pharma and MedTech.
- While first-time funding (angel and seed-stage deals) accounts for nearly one-third of the overall deal count, they represent only 6.4% in total volume (in 2020, it dropped to 5.5%).

---

[6]P. van der Velden, Lumira, Cost of MedDev in Medical Device PlayBook, Starfish Medical, Oct 27, 2020.

[7]Silicon Valley Bank (2021). Q4 2020 PitchBook NVCA Venture Monitor. Retrieved from https://www.svb.com/trends-insights/reports/venture-report/4q-2020-report

Table 6.1 Sectors of VC healthcare funding in the United States and Europe (in $M). *Data source*: Silicon Valley Bank[8]

Sectors	2018			2019			2020		
	US	Europe	Total	US	Europe	Total	US	Europe	Total
BioPharma	$15,147	$2,814	$17,961	$12,552	$3,132	$15,684	$19,968	$4,681	$24,649
HealthTech	$6,439	$575	$7,014	$7,091	$1,467	$8,558	$9,972	$1,382	$11,354
Dx/Tools	$5,391	$629	$6,020	$4,443	$940	$5,383	$8,708	$1,598	$10,306
Devices	$4,134	$609	$4,743	$3,953	$883	$4,836	$4,832	$569	$5,401
Total	$31,111	$4,627	$35,738	$28,039	$6,422	$34,461	$43,480	$8,230	$51,710

Table 6.2 MedTech in the context of VC healthcare funding in the United States and Europe (in $M). *Data source*: Silicon Valley Bank[9]

Sectors	2018			2019			2020		
	Total	Deals	Ave Deal	Total	Deals	Ave Deal	Total	Deals	Ave Deal
HealthTech	$7,014	463	$15.2	$8,558	564	$15.2	$11,354	615	$18.5
Dx/Tools	$6,020	234	$25.7	$5,383	263	$20.5	$10,306	312	$33.0
Devices	$4,743	262	$18.1	$4,836	277	$17.5	$5,401	316	$17.1

[8]Silicon Valley Bank (2021). Healthcare Investments & Exits Annual 2021 Report. Retrieved from https://www.svb.com/globalassets/library/managedassets/pdfs/healthcare-report-2021-annual.pdf
[9]Silicon Valley Bank (2021). Healthcare Investments & Exits Annual 2021 Report. Retrieved from https://www.svb.com/globalassets/library/managedassets/pdfs/healthcare-report-2021-annual.pdf

Here and after, we will use the following definitions: an early stage is defined as Series A and B, the late stage is defined as Series C and after.

6.3.1 Angel and Seed Stage

Angel and seed funding represents the first funding from professional investors. What is the difference between them? Oftentimes these terms are used interchangeably. However, some sources (e.g., footnote 10) separate them into two different buckets. We can presume that both buckets refer to seed funding. But, angel funding is provided by angel groups, while seed funding is provided by institutional investors other than angel groups.

What is seed funding? According to Investopedia,[11]

- "Seed capital is the money raised to begin developing an idea for a business or a new product.
- This funding generally covers only the costs of creating a proposal."

In the MedTech world, this definition seems entirely irrelevant. With an average pre-money valuation of $9.3M,[12] investors expect much more than just proposals. In general, it needs to be a working prototype or something similar. Thus, technically, the startup looking for seed funding needs to be at the end of the R&D phase.

Another definition of a seed-stage company is having "an established product/market fit." This definition seems more relevant, particularly for HealthTech ventures.

According to footnote 13, the seed funding started to significantly outperform angel funding recently. The average seed deal is around $2.4M, while the average angel deal is approximately $1.4M (numbers for all startups, not just MedTech). This trend is related

---

[10]Silicon Valley Bank (2021), Q4 2020 PitchBook NVCA Venture Monitor. Retrieved from https://www.svb.com/trends-insights/reports/venture-report/4q-2020-report

[11]Seed Capital, in Investopedia. Retrieved from https://www.investopedia.com/terms/s/seedcapital.asp, Accessed on Nov 8, 2020

[12]Silicon Valley Bank (2021), Q4 2020 PitchBook NVCA Venture Monitor. Retrieved from https://www.svb.com/trends-insights/reports/venture-report/4q-2020-report

[13]Silicon Valley Bank (2021), Q4 2020 PitchBook NVCA Venture Monitor. Retrieved from https://www.svb.com/trends-insights/reports/venture-report/3q-2020-report

to the institualization of seed rounds, which we mentioned before. The influx of prominent non-traditional players, which started participating in seed rounds (e.g., family offices), created significant upward pressure on the seed stage's average deal size.

However, angel/seed rounds still represent a small fraction of deals (in size) of all rounds of funding. As we mentioned already, although first-time funding accounts for nearly one-third of the overall deal count, they represent only 6.4% in total funding (in 2020, it dropped to 5.5%). Based on these numbers and taking into account Table 6.2, we can deduce that the average first funding deal size in MedTech is around $3.5M.

6.3.2   Early Stage

In Table 6.3, one can see Series A statistics for MedTech segments.

One can see that the numbers are in line with the average deal size for all startups, which has grown from $5M to $10M from 2010 to 2020.[14]

The definition of a Series A—ready startup is different for various MedTech sectors. HealthTech startups are quite similar to B2B and B2C startups. The typical requirement is to have a) a product on the market and b) substantial recurring revenue (e.g., $1–2M ARR for Series A and $5M for Series B).

However, for Diagnostics and MedDev, the criteria are obviously different. One possible criterion can be a complete elimination of technical risks.

6.3.3   Late Stage

In footnote 15, one can find the number of late-stage investment deals by number and total amount in 2010–2020. The overall deal value has grown 5× during this period, while the deal count nearly doubled. It resulted in the average deal size growth from $12M to $30M. The average pre-money valuation has reached $565.9M in 2010.

---

[14]Silicon Valley Bank (2021), Q4 2020 PitchBook NVCA Venture Monitor, Retrieved from https://www.svb.com/trends-insights/reports/venture-report/4q-2020-report

[15]Silicon Valley Bank (2021), Q4 2020 PitchBook NVCA Venture Monitor, Retrieved Apr 25, 2021 from https://www. svb.com/trends-insights/reports/venture-report/4q-2020-report

Table 6.3 Series A MedTech deals in the United States and Europe (in $M). *Data source:* Silicon Valley Bank[16]

Sectors	2018			2019			2020		
	Total	Deals	Ave Deal	Total	Deals	Ave Deal	Total	Deals	Ave Deal
HealthTech	$1,112	165	$6.7	$1,360	197	$6.9	$1,538	185	$8.3
Dx/Tools	$703	78	$9.0	$772	88	$8.8	$795	88	$9.0
Devices	$913	86	$10.6	$732	88	$8.3	$614	66	$9.3

[16]Silicon Valley Bank (2021), Healthcare Investments & Exits Annual 2021 Report, Retrieved from https://www.svb.com/globalassets/library/managedassets/pdfs/healthcare-report-2021-annual.pdf

6.4   Final Considerations

Multiple aspects should be kept in mind while considering raising money. However, it is a massive topic, and various books and articles are devoted to it. Instead, we will focus just on several practical considerations: the number of efforts, realistic timelines for the fundraising campaign, and geographical differences.

Firstly, founders should be aware of the number of efforts involved in raising funds. It is a long process (which will be discussed next), which requires very significant efforts and a systematic approach.

Sales professionals often talk about sales funnels and leads, which are being moved along the funnel. The same considerations can be applied here. As a startup executive, you are a salesman (or saleswoman), which sales opportunity to participate in the startup. Thus, the same concept of the funnel can be applied.

There are several stages of investor awareness:

- Investor leads—e.g., investors targeted during the fundraising campaign
- Qualified investor leads, where the dialog is already in progress
- Developed investor relationships, where the interest is already validated
- Allocations

This investor funnel and realistic fundraising campaign metrics are depicted in Fig. 6.2.

From Fig. 6.2, you can understand how significant this undertaking is. In order to raise funds, you need to approach 200–300 investors. And it is not just random contacts. It should be 200–300 investors, which are relevant to your startup and its current stage. In order to do it, you need to have grit. Making that many presentations is hard emotionally. It is also time-consuming. Thus, while raising money (which is an almost permanent state in startups), CEO needs to allocate most of their time to this process. Therefore, having other co-founders or founding team, which can back the CEO during this time, is very important.

Secondly, you have to be aware of the fundraising process's expected time (from start to closing the deal) and plan your activities accordingly.

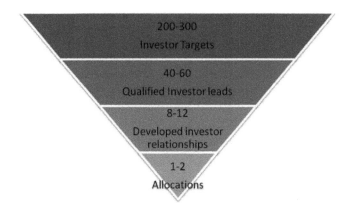

Figure 6.2 Typical funding allocation funnel according to footnote 17.

The fundraising process is quite long. On average, it takes at least 4–6 months. However, this time can be longer (e.g., during summer vacations) or significantly longer (e.g., more than a year when dealing with multiple angel investors). Even grant funding, in many cases, takes 4–6 months.

Thus, you need to consider these timelines and plan your activity accordingly. For example, you have to have a substantial runway to start the fundraising process. Investors may be uneasy if they learn that you have a six-month runway left.

Finally, there are significant geographical differences in resource availability and even startup valuation.

As we mentioned before, many investors (especially at very early stages) invest in "own backyard" companies. Thus, the availability of funds depends on how many investor groups are in a particular region. Moreover, regional investors may have or may not have a particular interest in life science in general and MedDev, in particular. Similarly, grant availability, subsidies, and tax incentives may vary significantly between even nearby regions.

The different compositions of investor pools and resource availability lead to significant geographical differences in startup valuations. For example, if $2–3 Mln is probably a lower 10% percentile valuation for a Silicon Valley's seed-stage company, the typical valuation for a seed-stage company in Canada is around $500,000.

---

[17]Ford D., *The Life Science Executive's Fundraising Manifesto*, Life Science Nation, 2014.

Chapter 7

IP and Other Moats

Middle Ages were quite opportunistic times. The vast Roman Empire was replaced by numerous fragmented states. Europe was segmented even further in a number of city-states. The thriving cities were under frequent attacks from the neighbors. Close to the coast, pirates were also an imminent threat.

The cities tried to protect themselves. Typically, the first layer of protection was natural features. For example, in Calabria, Southern Italy, the whole area was under constant attacks from Northern Africa. In order to protect their cities, people built them on top of hills.

High walls were the next layer of protection. Most medieval cities and castles had massive walls.

However, in the absence of natural features, a wall itself was not enough to protect a city or castle. It could be easily conquered using ladders. Thus, in the absence of natural defense, people built additional artificial fortifications, like moats.

A medieval castle moat was a deep and wide ditch surrounding castles for the purpose of defense. The moat is not a medieval invention. They were found in ancient Egypt, Africa (Benin City), Asia (China and Japan), and North America (Maya and Mississippian cultures).

Just in the 18th century, the fortification of cities and castles came to an end. With the proliferation of firearms, cities' and castles'

Bringing a Medical Device to the Market: A Scientist's Perspective
Gennadi Saiko
Copyright © 2022 Jenny Stanford Publishing Pte. Ltd.
ISBN 978-981-4968-25-6 (Hardcover), 978-1-003-31221-5 (eBook)
www.jennystanford.com

reinforcement lost its appeal and was reserved to certain military establishments like bastions.

Modern businesses also carry about the protection of their business. They deploy a series of measures to get protection from competition. In business jargon, these measures are often referred to as moats.

According to Investopedia,[1] "economic moat refers to a business's ability to maintain competitive advantages over its competitors in order to protect its long-term profits and market share from competing firms." The term was popularized by Warren Buffet.

There are multiple ways in which a company creates a structural deterrence that allows it to have an advantage over its competitors.

Cost Advantage

If a company has a significant cost advantage, it may deter any attempt to move into their industry by launching a price war and undercutting any competitor's prices. This move will either force the competitor to leave the industry or at least hamper its growth. Thus, companies with significant cost advantages are able to maintain a substantial market share of their respective industry by squeezing out any new entrant. Walmart is a famous example here. While such a scenario sounds great in theory, it can be challenging to achieve a sustainable cost advantage by itself. However, it can be feasible in combinations with other factors like patents, trade secrets, economy on a scale, etc.

Scale Advantage

While being big creates known inefficiencies, it can also create a sustainable competitive advantage through the economy of scale. Economies of scale refer to the case when more units of a good or service can be produced on a larger scale, which results in lower costs per unit. It also allows reducing overhead costs (e.g., administrative costs and advertising) and cost of capital. Therefore, large companies tend to dominate core market share in a particular industry. At the same time, smaller players are typically forced to either leave the industry or move into smaller "niche" segments. Unfortunately,

---

[1]Gallant C, What is an Economic Moat? Retrieved from https://www.investopedia.com/ask/answers/05/economicmoat.asp

achieving economies on the scale is hard to accomplish for early-stage startups before becoming the next Google or Facebook. More often, startups can be found on another side of the fence trying to disrupt a large player's business with economies of scale.

High Switching Costs

Being a big player in the industry has other advantages. Suppliers and customers of well-established companies can face high switching costs if they choose to do business with a new competitor. Thus, these burdensome switching costs impede carving market share away from the industry leader by any new entrants. It is particularly true for healthcare, where procurement is a long and rigorous process primarily due to multiple regulations. Therefore, switching costs are very high, and there should be particularly strong reasons for replacing an existing solution.

Intangibles

Intangible assets, like patents, brand recognition, regulatory approvals, government licenses, also can be an economic moat. For example, strong brand name recognition allows charging a premium for its products over other competitors, leading to higher profit margins (Apple, Louis Vuitton). However, this method works well for established companies. Strong intellectual property (IP) protection is a more practical strategy for startup companies to deter competition. Regulatory approval is another realistic way to create structural deterrence.

Above we presented just several examples to illustrate the concept. However, there are multiple other ways to achieve competitive advantages. In Fig. 7.1, one can see a pretty broad taxonomy of moats.

While moats are important for the company in general, it is of paramount importance for potential investors. From an investor's perspective, it is optimal to invest in a growing company that just started to harvest the benefits of broad and sustainable economic moats. Obviously, in this case, the longevity or sustainability of the moat becomes the most critical factor. The longer a company can reap profits, the better for itself and its shareholders. Therefore, creating sustainable moats is essential in making the company attractive to investors and ultimately in company success.

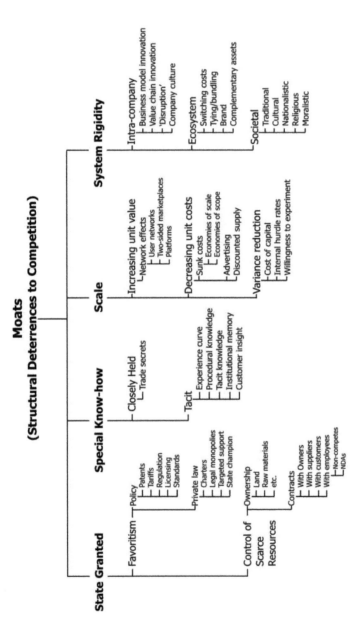

Figure 7.1 A taxonomy of structural deterrences to the competition. Reproduced with permission from footnote 2. Copyright, Reaction Wheel.

[2] J. Neumann, Reaction Wheel (2019). A taxonomy of moats, Retrieved from http://reactionwheel.net/2019/09/a-taxonomy-of-moats.html

Below we will consider several strategies, which can be of particular relevance to MedDev startups. Even though, according to Fig. 7.1, moats can be split into four big groups (state-granted, know-how, scale, and system rigidity), all our examples will belong mostly to sections of the first group (favoritism and control of scarce resources). For further examples, we refer to other excellent resources.[3]

We will start with intellectual property.

7.1   IP

Intellectual property (IP) is a broad class of intangible assets, bringing substantial competitive advantages to the company.

7.1.1   Types of IP

When people refer to IP, they typically mean "patents." Patents are the most publicized form of IP. However, it is not the only one. Patents are one of four pillars of intellectual property:

- Patents and Industrial Designs
- Copyrights
- Trade Secrets
- Trademarks

The typical IP protection strategy may contain some or all these elements. We will briefly consider all of them.

7.1.1.1   Patents

Patents are the core component of intellectual property. Patents cover new and useful inventions (product, composition, machine, process) or any new and useful improvement to an existing invention. Patents give protection for 20 years since an original filing (priority date). The idea behind a patent is to create a negative space where your competitors cannot operate. Your goal is to expand this negative space as much as you could.

---

[3]J. Neumann, Reaction Wheel (2019). A taxonomy of moats. Retrieved from http://reactionwheel.net/2019/09/a-taxonomy-of-moats.html

What are the main motivations behind patent filing? Here are several primary reasons behind it:

- It creates a negative space. It means that it can exclude competitors from the marketplace or give you a cost advantage over competitors.
- Even if your competitors managed to bypass your patent, your patent still gives you advantages. Namely, your competitors spent time and money bypassing the patent, not on product development or marketing.
- A patent portfolio looks impressive to investors. Even though, in most cases, they do not want their money spent on IP protection; however, they will appreciate an existing portfolio. Moreover, in many cases, the presence of patents can be an almost necessary requirement for investment.
- You can use your patent portfolio for cross-licensing with your competitors. In this case, it can give you some foothold in the competitive marketplace.
- You can use your portfolio to generate extra revenue by licensing it to other industries or geographical areas. For example, you plan to hold the EU market only and do not intend to directly go to the Chinese market. In this case, you can license your technology to another firm to bring it to the Chinese market.

In a nutshell, in some cases, patent filing's monetary benefits (i.e., an intrinsic value associated with the ability to generate or raise money, see main motivations above) can be higher than money spent on IP protection. In this case, filing a patent can be a valuable option.

There are several important aspects of the patent:

- The inventor/owner provides full public disclosure of the invention
- In exchange for this disclosure, the government grants exclusive rights to the owner for a set period of time (e.g., the United States or Canada—get a monopoly for 20 years from the date of filing)
- A patent is a right to stop others from making/using/selling the patented invention

- A patent is a country-specific right
- A patent is not a right to practice but to exclude others (negative space)

To get the idea patented, it needs to meet certain criteria. There are four widely accepted requirements for obtaining patent protection: novelty, non-obviousness, utility, and subject matter. While specific requirements may vary between countries, these core concepts are pretty universal. We will illustrate them using the US as an example with references to the US Patent Act, which is codified in Title 35 of the United States Code (USC):

1. Novelty (35 USC §102)

An examiner will determine whether the invention is new. To decide on novelty, an examiner compares the invention to prior known products or processes (referred to as the "prior art"). Prior art consists of all publicly available information in any form before a given date that might be relevant to a patent's claims of originality.

The inventor can provide some references about prior art by citing issued patents, published applications, scientific articles, technical papers, advertisements, and books to frame the invention's scope. However, an examiner can use other references to decide about novelty.

Novelty is one of the primary causes of claim rejections in patent applications. However, "novelty" has different meanings in different countries.

- Relative novelty – some jurisdictions, like Canada and the United States, have a 12-month grace period. This means that if you have disclosed your invention to the public (including sale or offer to sell), a 12-month grace period allows you to file the patent application, and the disclosure will not prevent registration. Note that the grace period is reserved for a disclosure made by the inventor only. Public disclosure by any other person before the filing voids the novelty.
- Absolute novelty – other jurisdictions, like the European Union, do not have the grace period. Any public disclosure, sale, or offer to sell before patent filing makes your patent not novel.

2. Not Obvious (35 USC §103)

An examiner will determine whether the invention is obvious. Obviousness is another primary cause of claim rejections in patent applications. The invention should be a non-obvious improvement over the prior art. This basis for such determination is non-obviousness "to a person having ordinary skill in the art to which the claimed invention pertains" at the time of filing. In plain words, the invention is compared to the prior art in order to determine whether the differences in the new invention would have been obvious to a person having ordinary skill in the type of technology used in the invention.[4] In order to make such determination, an examiner can combine multiple references or knowledge of a person skilled in the arts.

3. Utility or Usefulness

An examiner will determine whether the invention has a useful function. For medical devices or information technologies, the usefulness requirement typically can be easily met. However, the requirement becomes more important for Pharma and BioTech companies patenting a pharmaceutical or chemical compound. In this case, it is necessary to establish a practical or specific utility for the new compound.[5]

4. Subject Matter (35 USC §101)

Not all ideas can be patented. In general, the patentable and unpatentable subject matter can be described as follows:

- Patentable – "a product, a composition, apparatus, machine, process, or improvements thereto" [35 USC §102)].
- Unpatentable – Scientific principles and abstract theorems/formula, methods of medical treatment or surgery, processes that are exclusively mental steps.

Currently, all countries in the world use the "First-to-File" (FTF) approach. The United States was the last country to switch from the "First-to-Invent" (FTI) system, which took into account

[4]BitLaw, Patent Requirements. Retrieved from https://www.bitlaw.com/patent/requirements.html
[5]BitLaw, Patent Requirements. Retrieved from https://www.bitlaw.com/patent/requirements.html

an inventor's evidence of pre-filing conception, diligence, and reduction-to-practice. In 2013 by passing the Leahy–Smith America Invents Act (AIA), the United States adopted the "First-Inventor-to-File (FITF)" regime. The FITF system differs from FTF by allowing early disclosers some "grace" time before they need to file a patent (1 year in the United States). However, it should be noted that relying on such subtle differences may cause significant problems in other jurisdictions. For example, the European Patent Office (EPO) does not recognize any grace period, so early disclosure under the FITF provisions is an absolute bar to later EPO patents.

Patents do not provide IP protection cross-border. They need to be filed in each country where you are seeking IP protection. And each filing is quite costly. So, historically, an inventor had to spend a lot of money if they wanted to protect your invention. However, filing patents does not guarantee commercial success. Most of the patents filed have never been used by anybody and have not generated a single penny. Thus, by filing a patent application, you had to spend quite a significant upfront cost without any guaranteed return.

Fortunately, nowadays, other options are available. In 1970, the Patent Cooperation Treaty (PCT) had been signed. It allows seeking IP protection simultaneously in most countries (including all industrialized) by filing a so-called PCT application. It does not eliminate the necessity of national phase filings. However, it defers these high costs for 30 months. So, for a fraction of the cost of national phase filings, the PCT route gives you 2.5 years to decide whether it is worth incurring these costs and which markets are important for you.

If you follow the PCT route, your patent application will be published and become publicly available after 18 months since the original filing.

In addition to that, some countries (e.g., the United States and Canada) have a concept of "provisional patents."

Therefore, the patent process in North America (the United States and Canada) consists of three stages:

- Filing a provisional patent application
- Filing a PCT patent application
- Filing a national patent application

The first step consists of filing a provisional patent application. This filing is not examined and not disclosed to the public. Moreover, claims are not required. While it is not a patent, it gives 12 month period to file the PCT patent application. Note that provisional patent applications cannot be filed for the industrial designs, which will be considered in the next section.

The second phase is to file a PCT patent application. To keep the priority date, it should be filed within 12 months since the provisional patent filing. The PCT filing typically occurs through the patent office in the home country of the inventor. PCT filing is examined and gives protection for 18 months. It also got published six months after the filing. Thus, the invention can be kept secretive for 18 months (12 months provisional +6 months PCT). To keep the PCT filing priority, you can elaborate only on existing ideas filed in the provisional patent application. If you come up with new ideas, then they will be given a more recent priority date.

Finally, within 30 months (12 months provisional + 18 months PCT) since the original provisional patent application, you need to enter the national phase and file patent applications in each country where you are seeking protection. There are certain notes about this process. The national phase is where a thorough examination of your patent will occur. The examination is being conducted at the PCT stage as well; however, it is a light version of the examination. Passing the PCT review is a good sign; however, it does not guarantee to pass the national phase examination. National phase examination can be a long process and subject to much more scrutiny.

The second comment is that the national patent application needs to be filed in the native language of the country. Thus, if you wish to get protected in China or Japan, the patent needs to be translated into the respective language. Therefore, allocate some time and money for this translation.

These considerations can help to time the patent filing. Inventors need to consider the scope of claims (which may increase in time with new data and ideas) and their readiness to execute the plan.

Inventors have only 18 months before the idea becomes publicly available and just 30 months before they need to spend significant money during the national phase.

Of course, there is no necessity to file provisional patents or go the PCT route. The process can be streamlined and national patents filed at any time.

The communication with a patent office during the national phase is referred to as patent prosecution. It is a step where major spending (both time and money) may occur. The rules for the prosecution vary significantly between countries.

Below we list some country-specific requirements for the subject matter, national phase patent filings, and prosecutions.

Canada

Methods of medical treatment are not patentable in Canada. However, similarly worded "use" claims are allowable in Canada.

In Canada, the patent prosecution can be deferred. Examination of a Canadian patent application must be requested within four years from the filing date or, in the case of a Canadian national phase entry, four years from the international filing date.

Canada does not have a requirement to identify or file all known material prior art to the Canadian Patent Office (no Information Disclosure Statement). The applicant must only identify prior art cited in corresponding patent applications in other jurisdictions if requested by the examiner.

The maximum term of a Canadian patent is 20 years from the filing date of its application.

United States

In the United States, if the method meets the other patentability requirements (e.g., novelty and non-obviousness), medical treatment methods are patentable subject matter.

In the United States, the patent prosecution will start immediately upon filing a national application. The process will begin with an analysis of the patent for completeness. To be considered complete, the application must include: (1) a specification including at least one claim (see 37 CFR §§ 1.71–1.77), (2) an inventor's oath or declaration (see 37 CFR §§ 1.63–1.64), (3) requisite fees (see 37 CFR § 1.16), (4) and drawings, if necessary to convey the invention (see 37 CFR §§ 1.81–1.85).

Once it is considered complete, the patent will be forwarded to an examiner for substantive examination, where the prior art is examined.

Under United States patent law, while there is no duty to perform a search of the relevant art, "inventors and those associated with filing or prosecuting patent applications as defined in 37 CFR § 1.56 have a duty to disclose to the US Patent and Trademark Office (USPTO) all known prior art or other information that may be "material" in determining patentability."[6] This submission is called an Information Disclosure Statement (IDS).

The maximum term of a US patent is 20 years from the filing date of its application.

European Union

European patents are not to be granted in respect of "methods for treatment of the human or animal body by surgery or therapy and diagnostic methods practised on the human or animal body; this provision shall not apply to products, in particular substances or compositions, for use in any of these methods" [Article 53(c) EPC]. However, while a "method of treatment" is not patentable in Europe, a substance or composition may be patented for specific use in a method of surgery, therapy or diagnosis [Article 54(5) EPC].

In the European Union, applications can be filed for a national patent in each state or a single European patent application covering all member states. A European patent application does not result in a single patent. If the patent is granted, the application becomes a package of national patents, which will require validation in each country where the inventor seeks protection. Typically, validation consists of filing a translation into that state's national language and paying official fees to the national patent office.

In the European Union, the prosecution will start immediately upon filing a European patent application. The maximum term of a European patent is 20 years from the filing date of its application.

In addition to existing options, soon (tentatively: the beginning of 2022 as of the time of writing), inventors will be able to apply for a "Unitary Patent," which will be effective in nearly all European Union

---

[6]C. Sperry, E. M. Kingsland (2018). Information Disclosure Statements: When and How to File? Retrieved from https://www.mintz.com/insights-center/viewpoints/2231/2018-01-information-disclosure-statements-when-and-how-file

countries. The Unitary Patent will ensure uniform protection for an invention across the European Union and is expected to reduce costs and administrative burden.

7.1.1.2 Industrial designs

Industrial design (or Design Patent in the United States) focuses on protecting an article's visual features, namely its design, shape, pattern, or ornament. An industrial design protects a product's unique appearance, not what it is made of, how it is made, or how it works, which can be addressed by patents. Likewise, industrial designs do not protect intangible goods that are not visible.

An industrial design can only be registered if it is novel, meaning that the identical or substantially similar design has not been disclosed to the public. Some jurisdictions, like Canada and the United States, have a 12-month grace period. This means that if you have disclosed your design to the public, a 12-month grace period allows you to file the design, and the disclosure will not prevent registration.

Applications are kept confidential until registration or up to 30 months after their date of filing or priority. Once that confidentiality period is expired, the contents of the application will become available to the public. Thus, it is advisable to stage the registration of your design to ensure your product is ready to be launched at the same time.

There are two routes of industrial design registration:

- National route – you file Industrial Design applications in every country where do you want protection.
- International route – Hague System for the International Registration of Industrial Designs. In this case, you file one application and perform one payment in order to get protection in multiple countries, including the United States, the European Union, Canada, Japan, and South Korea. The Hague system is administered by the World Intellectual Property Organization (WIPO).

The term of protection for a Hague Registration is aligned with national terms. For example, in Canada, it begins on the date of its registration and ends on the later of the end of 10 years after registration in Canada or the end of 15 years after the date of International Registration.

Similar to patents, industrial design is a country-specific right. Each country has specific rules about patentability and the length of the protection.

Canada

In Canada, the industrial design registration gives protection (exclusive rights) for 10 years from the registration date or 15 years from the filing date, whichever period ends late.

United States

In the United States, a design patent is valid for 15 years from the date of issue, with no maintenance fees required to maintain the design patent.

European Union

In the European Union, industrial design is known as community design. The community design can be registered and unregistered. Protection for a registered community design (RCD) is for up to 25 years, subject to the payment of renewal fees every five years. Applications for designs can also be filed at the national or EU-wide level. A single application for a RCD at the EUIPO protects across the whole EU. An RCD is valid for five years and can be renewed up to four times for a total of 25 years.

7.1.1.3 Copyrights

Copyrights protect "original works of authorship:" original literary, artistic, dramatic, musical works, and other subject matter known as performers' performances, sound recordings, and communication signals. In the business context, they also cover computer programs.

In most countries, copyright is a form of protection provided to the authors of "original works of authorship" fixed in a tangible form of expression.

A work is "fixed" when it is captured (either by or under the authority of an author) in a sufficiently permanent medium such that the work can be perceived, reproduced, or communicated for more than a short time. Typically, copyright protection exists automatically from the moment the original work of authorship is fixed.

This standard is established internationally by the Berne Convention (1886), signed by most countries since then.

Copyright notice typically consists of the copyright symbol or the word "Copyright," the name of the copyright owner, and the first publication year.

In the United States, copyright law can be found in the Copyright Act, which is codified in Title 17 of the United States Code.

Copyright provides the owner of copyright with exclusive rights. For example, in the United States, an author has exclusive rights to:[7]

- "**Reproduce the work** in copies or phonorecords.
- **Prepare derivative works** based upon the work.
- **Distribute copies** or phonorecords of the work to the public **by sale or other transfer of ownership or by rental, lease, or lending.**
- Perform the work publicly if it is a literary, musical, dramatic, or choreographic work; a pantomime; or a motion picture or other audiovisual work.
- Display the work publicly if it is a literary, musical, dramatic, or choreographic work, a pantomime, or a pictorial, graphic, or sculptural work. This right also applies to the individual images of a motion picture or other audiovisual work.
- Perform the work publicly by means of a digital audio transmission if the work is a sound recording.

Copyright also provides the owner of copyright **the right to authorize others** to exercise these exclusive rights, subject to certain statutory limitations."

In the United States, the term of copyright is the author's life plus 70 years after the author's death. If the work is joint work of multiple authors, the period lasts 70 years after the last surviving author's death.

There is no such thing as an "international copyright" that will automatically protect an author's rights throughout the world. Instead, copyright is a country-based right. Protection against unauthorized use in a particular country depends on the national laws of that country. For example, in the United States, sections 104 and 104a of the US Copyright Act specify the scope of protection for unpublished and published foreign works.

---

[7]US Copyright Office, Copyright Basics. Circular 1. Retrieved from https://www.copyright.gov/circs/circ01.pdf

However, many countries offer protection to foreign works under certain conditions through international copyright treaties and conventions. Generally, a work may be protected in another country if that country has entered into an international agreement with the author's country. For example, the list of countries, which entered into agreements with the United States, can be found in Circular 38a[8] of the United States Copyright Office.

Transfer of Copyright Ownership

Copyright owner's exclusive rights (any, all, or parts of those rights) can be transferred. However, it is generally recommended to make the transfer in writing and signed by the rights owner.

In the US, a transfer of copyright ownership can be recorded with the Copyright Office. Even though the recording is not required to make a valid transfer between parties, it does provide certain legal benefits.

Copyright Registration

Copyright registration is often confused with the granting of copyright. Copyright in most countries nowadays is automatic on "fixation"—it applies automatically as soon as the work is fixed in some tangible medium. However, in addition to that, in some cases, copyrights can be registered with a government agency. For example, copyright registration was required in the United States before signing the Berne Convention in 1989.

Copyright registration can be considered as an extra layer of protection. For example, in the United States, registration offers several statutory advantages:[9]

- "Establishing a public record of a copyright claim.
- Before an infringement suit may be filed in court, registration (or refusal) is necessary for United States works.
- Registration establishes prima facie evidence of the validity of the copyright and facts stated in the certificate when registration is made before or within five years of publication.

[8]US Copyright Office, International Copyright Relationships of the United States, Circular 38a. Retrieved from https://www.copyright.gov/circs/circ38a.pdf
[9]US Copyright Office, Copyright Basics. Circular 1. Retrieved from https://www.copyright.gov/circs/circ01.pdf

- When registration is made prior to infringement or within three months after publication of a work, a copyright owner is eligible for statutory damages, attorneys' fees, and costs.
- Registration permits a copyright owner to establish a record with the US Customs and Border Protection (CBP) for protection against the importation of infringing copies."

Nowadays, in most countries, copyright registration is either voluntary or not required (no voluntary procedure available).

Canada

In Canada, copyright registration is voluntary. Copyrighted works can be registered at the Canadian Intellectual Property Office.

United States

In the United States, the copyright can be registered with the United States Copyright Office. Registration is required before filing an infringement suit in a US court. Moreover, copyright holders cannot claim statutory damages or attorney's fees unless the work was registered prior to infringement or within three months of publication.[10]

European Union

In the European Union, copyright registration is not required. Copyright protection in the European Union usually lasts for 70 years after the death of the author of the work. If the work originates outside the European Union and the author is not an European Union national, the term of protection granted in the European Union is, at minimum, 50 years after the death of the author of the work.

China

In China, copyright registration is voluntary. However, it is recommended, especially for software. Software Copyright Registration Certificate is a preliminary proof of registered items, which can help the holder to reduce the burden of proof in arbitration or lawsuit.

---

[10]US Copyright Office, Copyright Basics. Circular 1. Retrieved from https://www.copyright.gov/circs/circ01.pdf

7.1.1.4 Trade secrets

Trade secrets are a type of intellectual property that consists of certain information, expertise, or know-how that has been developed or acquired by firms. In a nutshell, a trade secret is any valuable business information, which is not generally known and is subject to reasonable efforts to preserve confidentiality.

A trade secret is protected from misuse by those who either obtain access through improper means, those who receive information from one who they know (or should have known) gained access through improper means, or those who breach a promise to keep the information confidential.

Trade secret laws exist in virtually any country. In common law countries (like the United States and Canada (except Quebec)), trade secrets are governed by the common law, ultimately derived from the English common law. While patents, copyrights, and trademarks are typically an exclusive area of federal legislation, trade secrets are an area of state/provincial legislation. Thus, historically, the law governing misappropriation of trade secrets developed separately in each state and province. Nowadays, the trade secrets laws are mostly unified across the United States. All states have adopted a portion or modified version of the Uniform Trade Secrets Act. According to footnote 11, "it allows trade secret owners to bring civil actions in federal court for trade secret misappropriation where the secrets relate to a product or service for use in interstate or foreign commerce."

Trade secret law varies from country to country. To standardize these practices and enable cross-border protection, certain international trade agreements contain trade secret clauses. For example, the former North American Free Trade Agreement (NAFTA) had provisions providing uniform minimum standards for protecting trade secrets.

Probably, the most publicized and enigmatic trade secret in the world is the original Coca-Cola recipe, which is kept secretive since 1891. It is known that Coca-Cola inventor John Pemberton shared his original formula with at least four people before his death in 1888. In 1891, Asa Candler purchased the rights to the formula from Pemberton's estate, founded The Coca-Cola Company, amended the

---

[11]Montaq, Protecting Trade Secrets In The U.S. and Canada. Retrieved from https://www.mondaq.com/canada/trade-secrets/503700/protecting-trade-secrets-in-the-us-and-canada

formula, and instituted the shroud of secrecy that has since enveloped the formula. The formula is kept in a vault, and presumably, not a single person knows all ingredients on the list.

This example demonstrates several important concepts. Firstly, the trade secret can be preserved for a long time if there are reasonable efforts to keep it confidential. Secondly, trade secrets can give a company substantial competitive advantages. Thirdly, that in order to claim trade secret protection, the company needs to instill some protective measures.

Trade secret or patent?

The choice between patent protection and trade secret protection requires careful consideration of several factors. In particular, it is important to consider the nature of the subject matter being protected:

- Can it be independently developed?
- Can it be reverse-engineered?
- Can it be maintained as a secret?
- How long will the subject matter have market value?
- Does the market value support investment in patent protection and enforcement?

7.1.1.5   Trademarks

A trademark is a sign or combination of signs used to distinguish their goods or services from others. Trademarks can be one or a combination of words, sounds, designs, tastes, colors, textures, scents, moving images, three-dimensional shapes, modes of packaging, or holograms.

Typically, a trademark is associated with an actual good or service (brand). However, over time, trademarks come to stand for not only the actual goods or services but also the reputation of the producer (a person or company). Therefore, trademarks can be very valuable intangible assets.

Trademark registration is country-specific. For example, by registering your trademark in Canada, you will gain exclusive rights to use it throughout Canada for 10 years.

Unlike other forms of IP, trademark registration is renewable. So, by paying fees and using the trademark properly, its registration can be perpetual. Another difference from other types of IP is the possibility of getting trademark registration revoked. Trademark registration can be canceled (expunged) for several reasons, including the trademark losing its distinctiveness, abandonment of the trademark, and non-use of the trademark.[12]

Not all signs or combinations of signs can be registered as a trademark. For example, in Canada, excluded items contain:

- "Names and surnames
- Clearly descriptive marks
- Deceptively misdescriptive marks—for example, you could not register "cane sugar" for candy sweetened with artificial sweetener
- Place of origin
- Words in other languages
- Confusing with registered or pending registration trademark
- Trademarks that are identical to, or can be mistaken for, prohibited trademarks"[13]—for example, official government design

The process of trademark registration is quite similar for most countries. Below we provide country-specific information using Canada as a base case.

Canada

To get the trademark registered, an applicant needs to file an application with a Trademarks Office (the Canadian Intellectual Property Office (CIPO) in Canada). The application must include a representation or description of the sign or both. It also needs to have a statement in specific and ordinary commercial terms of the goods and services associated with the trademark, including a product (service) class.

Upon receiving the application, the Trademarks Office reviews it for completeness. Once this process is finished, the Trademark Office acknowledges the application receipt and gives it a filing date.

---

[12]Canadian Intellectual Property Office, Trademarks Guide, Retrieved from https://www.ic.gc.ca/eic/site/cipointernet-internetopic.nsf/eng/h_wr02360.html
[13]Canadian Intellectual Property Office, What are Trademarks, Retrieved from http://www.ic.gc.ca/eic/siTe/cipointernet-internetopic.nsf/eng/wr03718.html

The next phase of the trademark registration process is application examination. As an example, we will continue with the examination process in Canada. The Trademarks Office (CIPO):

- searches the trademark database to find any registered or pending trademark that is confusing with the application,
- examines the application to make sure it does not contravene the trademarks acts and regulations, raise any objection to register your trademark, and inform you of any outstanding requirements,
- publishes the application in the Trademarks Journal, after which the public may file an opposition (challenge) to the application,
- registers the trademark if no one opposes the application (or if an opposition has been decided in the applicant favor).

In Canada, trademark registration is valid for 10 years since registration and can be re-registered subsequently long as the mark is in use in commerce.

United States

In the United States, trademark registration is valid for 10 years since registration and can be re-registered subsequently long as the mark is in use in commerce.

A quite detailed description of the trademark registration process in the United States can be found in the USPTO materials.[14]

European Union

In Europe, there are two primary paths to trademark registration:

- A European Union trademark (EUTM) provides protection in all member states of the Union.
- A national level trademark provides users with protection through the individual IP office of the EU country in which they apply.

The national trademark systems in each European Union state are often intertwined with EUTM. However, there are some discrepancies. There are pretty high barriers to an unregistered

[14]Basic Facts, USPTO. Retrieved from https://www.uspto.gov/sites/default/files/documents/BasicFacts.pdf

trademark's validity in the European Union and particularly in Germany. Unregistered trademarks enjoy protection under the German Trade Mark Act if they have acquired public recognition as a trademark by use in trade or constitute a well-known mark.

There is a notable difference in Germany's trademark registration process compared to EUTM, the United States, and Canada. In these systems, the trademark registration occurs after the opposition period. In Germany, the registration happens first, followed by a three-month opposition period.

The term of EUTM is 10 years and may be renewed every 10 years.

A trademark is a form of property. You can sell, donate, or transfer your rights to someone else through an assignment. To avoid ownership disagreements, it is recommended to notify the Trademarks Office about changes in ownership formally.

The trademark can be registered or unregistered. There are no requirements to register the trademark. By using an unregistered trademark for a certain length of time, you may have rights under common law (note that requirements for this may vary from country to country significantly). However, if you use an unregistered trademark and end up in a dispute, you could be looking at a long, expensive legal battle over who has the right to use it.[15]

The following symbols may designate a trademark:

- TM (the "trademark symbol," which is the letters "TM" in superscript, for an unregistered trademark, a mark used to promote or brand goods)
- SM (which is the letters "SM" in superscript, for an unregistered service mark, a mark used to promote or brand services)
- ® (the letter "R" surrounded by a circle, for a registered trademark)

The trademark registration can be done using:

- National application—applying in each country separately (including multiple languages)
- International application—the Madrid Protocol allows trademark protection in more than 100 countries by filing one single international application in one language with the World Intellectual Property Organization (WIPO).

---

[15]Pet Industry Law, Trademarks. Retrieved from http://petindustrylaw.com/trademarks.htm

The International Bureau checks only formal requirements to the application and does not conduct any substantive examination. If the international application meets all the applicable requirements, the International Bureau will register the trademark in the International Register.

Applications for registering a trademark using the Madrid system are examined according to each designated country's legislation and laws.

It is important to note that in order to file an international application, the applicant needs to file a national application in the native country (country of origin) first since a basic application or registration is required as a first step (referred to as a basic application or a basic registration).

Important Takeaways

- Trademarks can be very valuable intangible assets.
- Trademarks can be registered on unregistered. There are no requirements to register the trademark. By using a trademark for a certain length of time, you may have rights under common law. However, if you use an unregistered trademark and end up in a dispute, it can be a long and expensive legal battle.
- If you wish to keep your trademark registered, you must register your trademark the way you will use it. In other words, you must not change it in any way, including changing the color as you described it in your application.
- Trademark registration can be canceled (expunged) for several reasons, including the trademark losing its distinctiveness, abandonment of the trademark, and non-use of the trademark.

7.1.2 Considerations for Software

IP protection strategies may include multiple types of IP. A particularly relevant example of IP protection strategies can be software IP protection.

In general, the software can be protected using trade secrets, patents, and copyrights.

The idea (methods, algorithms) itself can be protected using patents or trade secrets. Patents may include ideas, systems, methods, algorithms, and embedded functions. For example, a programming algorithm or program language translation method can be patented.

However, the form, how this idea was implemented is protected by copyrights. For example, source code, object code, or machine code is covered by copyrights.

Therefore, patents and copyrights are complementary in IP protection for software. For example, let's assume a situation where software IP is protected by copyright only. The copyright covers the code's expression only and does not protect algorithms and methods. Thus, if the programming code can be reverse-engineered, then the software's innovation (concepts, features, or processes) can be extracted. Therefore, if the patent does not protect systems, algorithms, and methods, the infringer can bypass the whole IP protection by rewriting the code to avoid copyright infringement. Consequently, patents are particularly useful when the code can be reverse-engineered. In this case, patents and copyrights have been used together to protect the IP.

However, it should be noted that patent protection for software is much more challenging than for material ideas. The primary challenge here is to prove that the invention is not an abstract idea, which is not a patent-eligible subject matter. To become a patent-eligible subject matter, the invention incorporated in the software should not merely be an abstract idea but instead, have a technical character and include a technical contribution. Things are getting even more complicated in multi-country protection where different countries have different stances on the patentability of abstract ideas (e.g., the United States vs. European Union). However, the general rule of thumb is the more the software is linked with hardware or physical articles or machines such as a computer, the more likely it is to be considered by a patent office (and particularly by the USPTO in the United States) as having a technical character.[16]

If the code cannot be reverse-engineered, then the trade secrets can be helpful. Trade secrets are generally recommended to

[16]Al Tamimi, Types of IP Protection of Software Innovations. Retrieved from https://www.tamimi.com/law-update-articles/ip-protection-of-software-innovations

protect certain secret parts of a code, such as specific mathematic formulas, models, or secret recipes of the software that are not likely to be discovered by third parties through reverse engineering or independent discovery.[17] Google's search engine or Netflix's recommender system are good examples of software that can be protected by trade secrets (note that Google's original algorithm, PageRank, was patented instead).

7.1.3   Patent Infringement

The patentee has exclusive rights to produce, use, and sell the invention. For somebody else to use it legitimately, they need to obtain permission from the patent holder, which may usually be granted in the form of a license.

Patent infringement is a prohibited act with respect to a patented invention without permission from the patent holder. The definition of patent infringement may vary by jurisdiction, but it typically includes using or selling the patented invention.

A product infringes a claim of a patent if **every** element or limitation in a claim is present in the product, either literally or by equivalence

- The product may include more features than are included in the claim and still infringe the claim
- The product may look completely different than the patented invention and still infringe the claim

Infringement can be either direct or indirect.

Direct infringement

"Any act that interferes, in whole or in part, directly or indirectly, with the full enjoyment of the monopoly granted to the patentee during the term of the patent, without the patentee's consent, constitutes an infringement. Such an infringement is commonly referred to as a direct infringement."[18]

---

[17]Al Tamimi, Types of IP Protection of Software Innovations. Retrieved from https://www.tamimi.com/law-update-articles/ip-protection-of-software-innovations
[18]S. Garland, J. Want, D. Hnatchuk, M. Burt, Patent litigation in Canada: overview, Practical Law, Thompson Reuters

Indirect infringement

Indirect infringement is the inducement or procurement of another person to infringe a patent. However, proof of indirect infringement is typically subject to several tests, and the court will consider multiple factors. For example, in Canada, "the mere manufacture or sale of an article that is used by a second person in a way that infringes a patent has been found by courts to be insufficient to establish indirect infringement, even where the manufacturer/seller knew that the article would be used to infringe."[19]

In many countries, to amount to patent infringement, use is required to be commercial (or to have a commercial purpose). Several notable exceptions can be of particular importance for MedTech entrepreneurs. Firstly, in some countries (e.g., Canada), there is a common law experimental use exemption. Secondly, there is a regulatory use exception. For example, it is not an infringement to make, use or sell a patented invention solely for uses reasonably related to the development and submission of information required under any law of Canada.[20] Similarly, there are safe harbor provisions in the US to use a patented invention to gather data for regulatory submission.[21] Thirdly, in some countries (e.g., Canada), experimental use can be allowed under a statutory experimental use exemption, which is quite similar to a common law experimental use exemption.

7.1.3.1   Defenses to patent infringements

MedTech entrepreneurs can land on both sides of the IP protection fence. Their invention can be infringed, or they may inadvertently infringe somebody else invention. That is why it can be quite important to have a basic understanding of what is typically used as a defense in court. Below, we summarize several tactics, which is used in Canada:

Non-infringement

The product is missing an element recited in the claim.

---

[19]S. Garland, J. Want, D. Hnatchuk, M. Burt, Patent litigation in Canada: overview, Practical Law, Thompson Reuters.

[20]Lexology, In review: Patent legislation in Canada. Retrieved from https://www.lexology.com/library/detail.aspx?g=fe43954c-5576-4d69-a83f-01b033605772

[21]Sahu, P. K. and Mrksich, S. (2004) The Hatch-Waxman Act: When Is Research Exempt from Patent Infringement? ABA-IPL Newsletter 22(4).

Invalidity

The patent can be invalidated for several reasons, including prior art, obviousness, inutility, insufficiency of disclosure, etc.

Common law experimental use exemption

There would be a common law experimental use exemption if the patented invention were used in the course of not-for-profit experiments to determine if the patented article could be manufactured as set out in the patent or to develop an improvement on the patented article.

Regulatory use exemption

"It is not an infringement to make, construct, use or sell a patented invention solely for uses reasonably related to the development and submission of information required under any law of Canada or any country that regulates the manufacture, construction, use or sale of the product (Section 55(2), Patent Act). The regulatory use exemption can be raised in respect of activity related to compliance with the Canadian Food and Drug Regulations, including obtaining a Notice of Compliance from the Minister of Health for the sale of a pharmaceutical product in Canada."[22] A similar exemption exists in the United States.

Statutory experimental use exemption

In addition to the common law experimental use exception, some countries (e.g., Canada[23]) have a statutory exemption for the experimental use. In particular, under this exemption, an act committed for the purpose of experimentation relating to the subject matter of a patent is not an infringement of the patent.

Limitation period

The patent owner waited too long to enforce the patent. In some situations, where the patentee has delayed the enforcement of its rights, the defenses can be based on statutory limitation periods, as

---

[22]S. Garland, J. Want, D. Hnatchuk, M. Burt, Patent litigation in Canada: overview, Practical Law, Thompson Reuters.
[23]Subsection 55.3(1) of the Patent Act.

well as the common law defenses of acquiescence, delay, laches, and estoppel.[24]

For example, in Canada, the Patent Act provides a limitation period of six years for patents. Consequently, a patentee will not recover damages or profits for infringements that occurred more than six years before the start of the infringement proceeding.

It should be noted that the patent infringement litigation practices may vary significantly even within a single jurisdiction, particularly in common law countries. For example, a small county in Texas recently became a patent infringement litigation capital of the United States.[25]

7.1.4 Patent Search

In order to avoid infringing somebody else invention inadvertently, it can be helpful to understand your own IP position. To understand your IP position, you can perform a freedom-to-operate (FTO) search. FTO's goal is to assess your ability to produce and market products without legal liabilities to third parties.

FTO can be performed internally (in this case, it is typically called a preliminary patent search) or by a patent lawyer. In the latter case, you can additionally request an FTO opinion.

There are multiple pros and cons of obtaining the FTO opinion. It will definitely strengthen your case when you talk with investors. However, the biggest shortcoming of any significant effort in the FTO is its potential incompleteness. As we mentioned earlier, under the PCT route, the patent application is not publicly available for the first 18 months. Thus, newly filed applications will not be captured in an FTO or preliminary search.

Another nuance of FTO is that in the United States, all information gathered during FTO is required to be submitted to the patent office as an Information Disclosure Statement (IDS). This requirement will be discussed briefly at the end of this section.

---

[24]S. Garland, J. Want, D. Hnatchuk, M. Burt, Patent litigation in Canada: overview, Practical Law, Thompson Reuters.

[25]Artz K. (2020). Surprise—Waco, Texas, Is the Patent Litigation Capital of the United States! Retrieved from https://www.law.com/texaslawyer/2020/10/08/surprise-waco-texas-is-the-patent-litigation-capital-of-the-united-states

However, despite these shortcomings, there are still multiple good reasons to perform the preliminary patent search internally. Firstly, you may find some competitors. Secondly, you will see whether you can craft your claims to bypass them if you find them. Finally, the patent language is quite particular, so it is good to start getting familiar with it.

The preliminary search can be as simple as a search using Google Patents or free online resources of the World Intellectual Property Organization (WIPO), United States Patent and Trademark Office (USPTO), Canadian Intellectual Property Office (CIPO), or the European Patent Office (EPO).

The search can be performed using multiple fields. Several approaches can be utilized.

Competitors

Think about potential competitors operating in this space (e.g., potential owners/applicants/assignees).

Keyword search

Keyword search is one of the most helpful approaches. However, there are certain limitations. Firstly, the patent language differs significantly from everyday language. For example, "fastening device" can vary from zipper to shoelace to industrial screw to implant in dermatology. Secondly, words may have multiple alternative meanings (e.g., cloud vs. internet cloud), giving a significant number of false-negative results you need to sift through. Finally, due to the rapid pace of innovation, some words can be obsolete (e.g., CD-ROM); however, they still can be found in valid patents.

Thus, one can use a combination of search techniques to define a classification of the invention. Defining proper patent classification in combination with keyword search may significantly narrow down and simplify the search.

However, you should be aware that there are multiple classification systems (See Table 7.1). Some countries use several classification systems concurrently.

Table 7.1 Patent Classification Systems

Country	Classification Systems
US	US Patent Classification (USPC) and Co-operative Patent Classification (CPC) systems* *CPC: co-developed with the European Patent Office to replace the USPC
EU	International Patent Classification (IPC) system and Co-operative Patent Classification (CPC) systems
Canada	International Patent Classification (IPC) system and Canadian Patent Classification system

Country-specific requirements:

United States

Under the US patent law, while there is no duty to perform a search of the relevant art, "inventors and those associated with filing or prosecuting patent applications as defined in 37 CFR § 1.56 have a duty to disclose to the US Patent and Trademark Office (USPTO) all known prior art or other information that may be "material" in determining patentability."[26] This submission is called an Information Disclosure Statement (IDS).

Although there is no simple rule as to what information is "material," a good rule of thumb[27] is to disclose all information that is relevant to the claimed subject matter, including other related US patent applications and patents of the applicant, references cited in a PCT or foreign counterpart application, foreign references and non-patent literature.

According to footnote 28, while an IDS can be submitted at various stages of prosecution, "it is recommended to file an IDS

---

[26]C. Sperry, E. M. Kingsland (2018). Information Disclosure Statements: When and How to File? Retrieved from. https://www.mintz.com/insights-center/viewpoints/2231/2018-01-information-disclosure-statements-when-and-how-file

[27]C. Sperry, E. M. Kingsland (2018). Information Disclosure Statements: When and How to File? Retrieved from https://www.mintz.com/insights-center/viewpoints/2231/2018-01-information-disclosure-statements-when-and-how-file

[28]C. Sperry, E. M. Kingsland (2018). Information Disclosure Statements: When and How to File? Retrieved from https://www.mintz.com/insights-center/viewpoints/2231/2018-01-information-disclosure-statements-when-and-how-file

(and supplemental IDSs) as early as possible to ensure that it will be considered by the examiner and avoid incurring additional costs."

7.2 Regulations

As we discussed already in Chapter 2 (Regulatory Environment), the necessity to comply with various regulations makes MedTech projects much more complex and costly than similar hardware projects. However, the same applies to your potential competitors.

There are several competitive advantages associated with regulatory hurdles. We will discuss them briefly:

7.2.1 Time

Regulations give you a substantial handicap over "me too" competitors. As discussed in Chapter 5 (Timelines and Capital), the time required to bring a medical device to market is substantial for Class II and III devices. Thus, if a competitor wants to start a project from scratch and bring a similar device to the market, it will require 6–7 years for a Class II device and 10 years for a Class III device. Even assuming that the competitor started the project earlier than you brought your product to the market, it still gives you a several-year advantage over them.

This advantage can be translated into customer adoption, which in combination with high switching costs in healthcare, can lead to sustainable competitive advantage.

7.2.2 Money

The same principles apply to the costs of the project. As discussed in Chapter 5 (Timelines and Capital), the money required to bring a medical device to market is substantial for Class II and III devices. It is around $30 Mln for a Class II device and $90-100 Mln for a Class III device. This money can be affordable for a large corporation. However, it is prohibitively expensive for any startup. Investors will be very hesitant to give money to a "me too" competitor, which is several years behind the first mover. Thus, the probability of failure of any small competitor is very high.

A big corporation is still can be a threat. However, it is much easier for big companies to acquire a small competitor than compete with them. Moreover, the price tag ($100 Mln and up) is in the range of their acquisition targets.

That is the reason why investors prefer Class III devices. They give the startup (and investors) significant competitive advantages, which can frighten any potential competitor (its investors, primarily), thus creating a sustainable competitive advantage.

7.3   Data

Data can be another "unfair competitive advantage."

There are various ways how to monetize data. In typical data play, it is assumed that the company collects customer data, which later can be sold to multiple parties.

Another opportunity is to use customer data to refine advertisement, e.g., through tailored content. These strategies are used in social media, for example, by Twitter or Facebook. The advertiser can refine the target audience based on the required demographics (e.g., age, sex, education, income, occupation, location) and display its advertisement only to the target audience.

Google uses all available data (e.g., searches by user) to create a user profile and improve tailored content.

However, most of these strategies are not applicable in the MedTech world. Such as the MedTech industry is heavily regulated, there are multiple privacy rules, which prohibit unauthorized use of consumers' or patients' data.

United States

The United States does not have a federal-level general consumer data privacy law or a data security law. However, there are consumer data privacy regulations in various states (e.g., California Consumer Privacy Act in California) and strict federal regulations about patients' data. In the US, the Health Insurance Portability and Accountability Act of 1996 (HIPAA) is a federal law that required the creation of national standards to protect sensitive patient health information from being disclosed without the patient's consent or knowledge. This information is referred to as protected health information (PHI).

The HIPAA Security Rule demands that safeguards be implemented to ensure confidentiality, integrity, and PHI availability. At the same time, the HIPAA Privacy Rule places limits on the uses and disclosures of PHI.

PHI is any health information that can be tied to an individual. According to HIPAA, PHI includes one or more of the following 18 identifiers.

1. Names (Full or last name and initial)
2. All geographical identifiers smaller than a state, except for the initial three digits of a zip code if, according to the current publicly available data from the US Bureau of the Census: the geographic unit formed by combining all zip codes with the same three initial digits contains more than 20,000 people, and the initial three digits of a zip code for all such geographic units containing 20,000 or fewer people is changed to 000
3. Dates (other than year) directly related to an individual
4. Phone Numbers
5. Fax numbers
6. Email addresses
7. Social Security numbers
8. Medical record numbers
9. Health insurance beneficiary numbers
10. Account numbers
11. Certificate/license numbers
12. Vehicle identifiers (including serial numbers and license plate numbers)
13. Device identifiers and serial numbers
14. Web Uniform Resource Locators (URLs)
15. Internet Protocol (IP) address numbers
16. Biometric identifiers, including finger, retinal, and voice prints
17. Full face photographic images and any comparable images
18. Any other unique identifying number, characteristic, or code except the unique code assigned by the investigator to code the data

If these identifiers are removed, the information is considered de-identified protected health information and not subject to the HIPAA Privacy Rule's restrictions.

PHI can be used to provide a healthcare service, such as a diagnosis or treatment or billing. However, before using PHI, the company needs to obtain consent from the patient for all other uses.

Canada

In Canada, the Personal Information Protection and Electronic Documents Act (PIPEDA) is the federal privacy law for private-sector organizations. It sets out the rules for businesses to handle personal information in the course of their commercial activity.

Under PIPEDA, personal information includes any factual or subjective information, recorded or not, about an identifiable individual. This includes information in any form, such as:

1. age, name, ID numbers, income, ethnic origin, or blood type;
2. opinions, evaluations, comments, social status, or disciplinary actions; and
3. employee files, credit records, loan records, medical records, the existence of a dispute between a consumer and a merchant, intentions (for example, to acquire goods or services or change jobs).

Organizations covered by PIPEDA must obtain an individual's consent when they collect, use, or disclose their personal information. Personal information can only be used for the purposes for which it was collected. If a company is going to use it for another purpose, they must obtain consent again.

To make things even more complicated, PIPEDA does not apply to organizations that operate entirely within Alberta, British Columbia, or Quebec. These three provinces have general private-sector laws that have been deemed substantially similar to PIPEDA.

European Union

In the European Union, General Data Protection Regulation (or so-called GDPR rule) is a regulation that requires businesses to protect the personal data and privacy of European Union citizens for transactions that occur within the European Union. GDPR is considered the world's strongest set of data protection rules. The regulation became a model for many national laws outside the European Union, including Japan, Brazil, and South Korea.

Here we list three of the most relevant GDPR requirements out of 10 discussed in footnote 29:

- "Transparency—companies must inform data subjects about the processing activities on their personal data
- Limitation of purpose, data, and storage—the companies are expected to limit the processing, collect only necessary data, and not keep personal information once the processing purpose is completed. This would effectively bring the following requirements:
  - forbid processing of personal data outside the legitimate purpose for which the personal data was collected
  - mandate that no personal data, other than what is necessary, be requested
  - ask that personal data should be deleted once the legitimate purpose for which it was collected is fulfilled
- Consent—if the company intends to process personal data beyond the legitimate purpose for which data was collected, clear and explicit consent must be asked from the data subject. Once collected, this consent must be documented, and the data subject is allowed to withdraw his consent at any moment".

Any violation of privacy laws brings harsh civil penalties (e.g., it may expose unlimited liability). Thus, any use of personal or medical data needs to be carefully crafted by taking into account multiple regulations in various jurisdictions. The proper agreements need to be put in place with care providers, and consent may need to be obtained from patients.

Therefore, the primary use of data in the MedTech world is to utilize this data internally to refine algorithms or develop new functionality based on insights, which can be derived from data.

This will allow the first mover to stay ahead of the game and deliver value to customers, which cannot be delivered by somebody else who lacks this data.

---

[29]Bhatia P., A summary of 10 key GDPR requirements. Retrieved from https://advisera. com/eugdpracademy/knowledgebase/a-summary-of-10-key-gdpr-requirements/

7.4 Conclusions

Creating a sustainable competitive advantage can be a multi-stage process when different individual methods interplay and reinforce each other. For example, regulatory approval is a significant hurdle for any new entrant, which will defer any competition for several years. That time period can be used to achieve traction, which can lead to high switching costs (system rigidity) or creating own data, which will drive further competitive advantages.

However, it should be noted that economic moats are generally difficult to pinpoint at the time they are being created.[30] Their effects are much more easily observed in hindsight once a company became successful. In many cases, it is a combination of luck and the right timing. However, having a systemic approach and plan is still a good idea.

---

[30]Gallant C., What Is an Economic Moat? Investopedia. Retrieved from https://www.investopedia.com/ask/answers/05/economicmoat.asp

Chapter 8

Business Model

What is a business model? If you try to answer this question, you may find that while it is easy to grasp intuitively, it is not easy to formulate it rigorously. You are not alone. There is no well-established description. Moreover, the definition of the "business model" evolved in time, and you might be surprised to hear that it is a fairly new concept traced back to 1994. A good overview of the evolution of the "business model" concept is presented elsewhere.[1]

While the "business model" can be a quite philosophical term in general, it has a very practical meaning in the startup world. That is because investors want to know how the startup will make money.

In a nutshell, the business model is a method of how the business can **repeatedly** sell its product or service. As you may notice, the keyword here is "repeatedly." While revenue is good in general for the business's bottom line, the quality of revenue can vary from investors' perspective.

Firstly, revenue can be split into operating vs. non-operating revenue. Operating revenue is the revenue that a company generates from its primary business activities. Everything else is non-operating revenue. Selling medical devices, which you developed and manufactured, is operating revenue. Selling of company core assets or investment income are examples of non-operating revenue.

---

[1]Ovans, A. (2015). What Is a Business Model? *Harvard Bus. Rev*. Retrieved from https://hbr.org/2015/01/what-is-a-business-model

Bringing a Medical Device to the Market: A Scientist's Perspective
Gennadi Saiko
Copyright © 2022 Jenny Stanford Publishing Pte. Ltd.
ISBN 978-981-4968-25-6 (Hardcover), 978-1-003-31221-5 (eBook)
www.jennystanford.com

Obviously, operating revenue has much higher quality than non-operating one.

Operating income is typically further split into several buckets. For example, it can be divided into one-time and recurrent revenue. If you manufacture and sell the device for lab tests and this device will be used for three years, it will be one-time revenue. However, if you sell test kits for this device, it will be recurrent revenue. Subscriptions and commission/fee from every transaction or use are other examples of recurrent revenue.

The recurrent revenue has the highest quality, at least from the investor's perspective. It should be mentioned that there is nothing wrong with one-time, project-based revenue. One company can make $100Mln by doing $500k every day, while the other one does $100Mln by doing 5 large deals. To both companies, the OTHER model seems like a nice addition to the existing revenue. So, project-based revenue can be very good for the company's bottom line. However, they may be treated differently from the growth perspective.

Even if we talk about recurrent revenue, there are several additional important considerations.

Firstly, it is revenue continuity. Long-term customer contracts provide future revenue predictability. Thus, if your product is deeply integrated into customer's workflow and the switching costs (see Chapter 7) are high, then this recurring revenue is usually relatively stable.

Secondly, it is revenue diversity. The higher the customer concentration is, the greater the risk. If you have only one customer and your relationship is getting sour, it can be disastrous for your business. Early-stage companies tend to have only a few customers who make up a large portion of revenues, but they must strive to build a diverse revenue base over time. Ideally, no customer generates more than 10% of the revenue.

The two primary types of business models available to entrepreneurs are business-to-consumer (B2C) and business-to-business (B2B) models.[2] However, it is evident that such granularity is not sufficient to convince investors. Thus, a more elaborative approach is required.

---

[2]In reality this separation is blurred. According to Marc Benioff, the founder of Salesforce "We really see every B2B company and every B2C company becoming a B2B2C company."

There are several frameworks to consider while developing the business model.

8.1   Lean Canvas

The Lean Canvas is among such frameworks. While we try to avoid general startup topics as they are well covered in numerous other resources, the Lean Canvas concept is worth brief mentioning.

In recent years, the Lean Canvas became almost a universal tool for the business model's development and communication. For example, it nearly completely replaced a business plan as the method to converse with startup investors. The primary advantage of the Lean Canvas is its succinctness. Nobody is interested in reading a 40-page document; however, people will be willing to try comprehending a familiarly structured single-page diagram.

But even more importantly, the Lean Canvas is a helpful tool to conceptualize and refine the idea within the team. While creating a full business plan seems like a formidable and daunting task, the one-pager can be drafted in a matter of hours and perceived as a less intimidating undertaking. Moreover, it facilitates and accelerates communication within the team. Any inconsistencies and areas for improvement are clearly visible.

Make no mistake, writing a good Lean Canvas is a long continuous process. Writing a good short piece is much harder than just dump info in a large one. As Mark Twain once said, "I didn't have time to write you a short letter, so I wrote you a long one." Most likely, it will require multiple iterations, and all hypotheses need to be validated. However, due to its succinct form, it is perceived as a much less frightening task.

According to Wikipedia,[3] the Lean Canvas was developed in 2010 by Ash Maurya[4] as an adaptation of a Business Model Canvas (introduced in 2008 by Alexander Osterwalder[5]) to a startup environment.

[3]Lean Startup. Retrieved from https://en.wikipedia.org/wiki/Lean_startup
[4]Maurya, A. (2012). Running lean: iterate from plan A to a plan that works. The lean series (2nd ed.). Sebastopol, CA: O'Reilly.
[5]Osterwalder, A.; Pigneur, Y.; Clark, T. (2010). Business model generation: a handbook for visionaries, game changers, and challengers. Hoboken, NJ: John Wiley & Sons.

The Lean Canvas consists of nine sections or blocks arranged graphically. There are several versions of this tool, including the original Business Model Canvas. However, the differences between them are minor.

The starting point in building the Lean Canvas is to identify customers. With your solution in mind, you can start brainstorming users, which can benefit from using your product. However, notice a difference between customers and users. The customer is someone that pays for your product. So, if you have multiple user roles, the ultimate goal is to identify customers.

It should be noted that the Lean Canvas model assumes experimental learning. We briefly discussed this approach in Chapter 4 (PDP). It involves applying a scientific method: building hypotheses based on experimental data and testing/validating them. Thus, extensive experimentation is required.

The four major steps here are design experiments to collect initial data, run experiments to collect this data, formulate a hypothesis based on this data, and test/ validate it. If the data does not support the theory, it needs to be discarded, and a new hypothesis formulated and tested.

Almost all sections of the Lean Canvas will require applying the scientific method and experimental verifications. However, different blocks of the Lean Canvas require different experimental techniques. For example, customer interviews will be necessary to understand a customer problem. However, verification of channels will require a completely different approach. In this case, for example, advertisements can be run on various social media platforms, and data on their performance (e.g., A/B testing) collected. Details of experimentation techniques used in the Lean Startup model can be found elsewhere.[6]

The second comment will be about the iterative nature of the Lean Canvas. The Lean Canvas is a live document that needs to be revised and updated upon new information arrival. The team needs to start developing the Lean Canvas very early. In particular, it can be an essential tool to establish whether the business case exists in the first place. At this stage, the Lean Canvas can be very rough. With

---

[6]Maurya, A. (2012). Running lean: iterate from plan A to a plan that works. The lean series (2nd ed.). Sebastopol, CA: O'Reilly.

customer interviews and other experimentation, it will become more granular and precise. It can be instrumental in decision-making support at various stages of the product development process (see Chapter 4).

Below we briefly discuss sections of the Lean Canvas arranged in a logical order how to fill them out. For further reading, we refer to an excellent book[7] by the inventor of the Lean Canvas.

8.1.1 Customer Segments

The possible customers need to be broken down into smaller, more homogeneous groups or customer segments. Each customer segment may have a unique set of problems. Thus, unique value propositions and channels need to be tailored to them. Consequently, a separate Lean Canvas needs to be developed for **each** customer segment. A good starting point is to create lean canvases for 2–3 primary customer segments.

However, the customers will still be heterogeneous even within each segment. Some of them are eager to try a new product. Others will defer the purchase until it gets widely adopted. Therefore, it is particularly important to identify potential early adopters within a customer segment.

8.1.2 Problem

For each customer segment, you need to identify their problem. If you are not from a particular customer segment, it needs to be done through customer interviews. Theorizing about somebody else problem can lead to very costly mistakes or even set the whole venture in the wrong direction.

After interviewing even several customers, you will start noticing recurring themes.

Again, it is imperative to identify a problem, which is large and real. You don't want to spend many years of your life tailoring an elegant solution to a non-existent problem or solving the problem for a non-existent or minuscule market.

---

[7]Maurya, A. (2012). Running lean: iterate from plan A to a plan that works. The lean series (2nd ed.). Sebastopol, CA: O'Reilly.

8.1.3   Unique Value Proposition

According to Steve Blanc, "Unique Value Proposition: A single, clear compelling message that states why you are different and worth buying."[8] It is beneficial to conceptualize your idea in several sentences. Unique Value Proposition (UVP) can be used as an elevator pitch or grab attention on your website.

Like most other parts of the Lean Canvas, UVP needs to be validated. You have to try it on potential customers and see whether it resonates with them.

UVP will be used intensively and drive the messaging going forward. Thus, it is worth spending time on writing and refining UVP. There are many templates available. You can find some of them in.[9]

8.1.4   Solution

This section is typically being built in iterations. As we mentioned, the Lean Canvas is a live document. It needs to be updated with the arrival of new information. For example, in the beginning, your hypotheses about the product are not tested. So, it is not worth spending much time hypothesizing about the final solution at this stage. Instead, it is better to sketch the proposed solution's main features and benefits next to each problem. Once you tested your hypotheses with users, you will be better positioned to revisit this section.

8.1.5   Channels

This section describes how you will get customers. There are multiple ways to do it. Moreover, these methods will evolve in time.

For example, for many MedTech entrepreneurs, participation in scientific and industry conferences and trade shows is one of the early primary channels to get customers. However, as the company scales and builds its own direct sales force, this channel most likely becomes secondary.

---

[8]Blank, S. (2020). The Four Steps to the Epiphany, Wiley.
[9]Retrieved from https://www.smartsheet.com/value-proposition-positioning-templates

The other relatively common approach is to partner with a larger industry player and use its distribution channels.

The "channels" section can be one of the most critical ones in the Lean Canvas. It shows immediately how mature and reasonable your business model is.

8.1.6 Cost Structure

It should include all factors, including customer acquisition (channels) and their support. It is quite difficult to predict your ultimate cost structure. Moreover, your cost structure will evolve in time. Thus, at the early stage, it is not worth digging deep into this topic. Back of envelope calculations and speaking with other founders will be the most helpful.

One rule of thumb for building a successful business is to remember that your customers' lifetime value should exceed customer acquisition cost by at least a factor of three.

8.1.7 Revenue Streams

The ability to generate revenue is the essence of any business. In general, there are four primary sources of revenue (or revenue streams): recurring revenue, transaction-based revenue, project revenue, and service revenue. However, this segmentation is very high level and not very helpful to synthesize your revenue streams. A more practical approach would be to look at specific examples around. Multiple new methods to generate revenue emerged in recent years, particularly with the advent of smartphones.

There are several points to mention here. Firstly, as we said before, the quality of revenue is essential, particularly for investors. Thus, the recurring revenue is of particular importance. Secondly, it is better to have several revenue streams instead of one. Thirdly, revenue streams (e.g., the price itself) also require experimentation.

Pricing

Pricing can be a potent tool in your armory. You can actually go as far as making zero changes to your device, changing a pricing model, and re-launching your product as a completely new one.

Different price models may have a different appeal to your customer segments. For example, procurement of medical devices above the certain price tag (typically $10,000) may require budget approval, which can significantly complicate and lengthen the sale cycle.

Pricing can be used to align the interests of both the client and the company. Moreover, different pricing models may reinforce or inhibit a particular customer behavior. For example, the one-time sale does not stimulate device usage. If there are no incentives to use the device more often, it may end up on the shelf, and your ability to collect usage data may be impeded.

Also, pricing needs to be aligned internally on sales compensation. Different pricing model incentivizes sales differently. Thus, it can be an essential part of alignment within the company.

Finally, an incorrect pricing model may drive your company out of business. Assume that you use a one-time sale model for your medical device. If you were successful and saturated the market with your device, you may suddenly experience a significant drop in sales and revenue. Customers may be satisfied with existing features/performance, continue to use your devices, and refuse to upgrade them. All this may result in a significant drop in sales, which may drive your company out of business.

If you developed a price model for your product and calculated its revenue projections, it can perform the following exercise. Switch the price model (e.g., from the one-time sale to subscription or from subscription to SaaS) and try to match your initial model's revenue projections.

If you have not developed the pricing model yet, it can be helpful to create three completely different pricing models.

The very helpful pricing model ideas pertinent to the Web space can be found elsewhere.[10]

8.1.8 Key Metrics

Your progress to goals needs to be measurable. Thus, it is important to develop a quantifiable approach. It can be split into two parts.

---

[10]C. Janz (2014). Five ways to build a $100 million business. Retrieved from http://christophjanz.blogspot.com/2014/10/five-ways-to-build-100-million-business.html

Firstly, you need to identify key activities. Then, for these activities, you can identify measurable metrics.

Note that as with most parts of the Lean Canvas, key metrics will evolve in time. For example, one of the core activities for a just-launched startup will be to run experiments to validate the hypothesis. Thus, the number of customer interviews can be a key metric here. However, once the product is launched, the set of activities and metrics will be completely different. For example, monthly recurring revenue (MRR) or customer retention become metrics very important to investors.

One important pitfall to avoid when setting metrics is the so-called "vanity metrics." These are measurements that do not materially impact your business but may seem relevant and flashy. They tend to be easily manipulated and not provide actual business value.

8.1.9 Unfair Advantage

According to Jason Cohen, everything worth copying will be copied. Thus, "a real unfair advantage is something that cannot be easily copied or bought."[11] You have to think well about how you will protect your business. It is of particular importance for investors. And again, it shows how reasonable and mature your business model is.

We discussed competitive advantages in the previous chapter (IP and Other Moats).

Now, when we have a reasonably good understanding of developing and validating business models, we can switch our attention to different types of business models.

8.2 B2C Model

While the B2C model is prevalent in the consumer goods market, it does not have the same relevance in the MedDev world. Again, regulations are the primary reason for that.

Just a tiny fraction of medical devices can be marketed or sold over-the-counter. For example, in many markets (e.g., the United

---

[11]Jason Cohen, A Smart Bear. Retrieved from http://blog.asmartbear.com/jason-cohen

States and Canada), the ECG cannot be sold over-the-counter. In the United States, the ECG is considered in FDA classification as a "monitor, cardiac (including cardiotachometer and rate alarm)" device with DRT product code. It is a Class II device, which requires a 510(k) premarket notification. It can be sold to a patient by prescription only.

If the device cannot be sold over-the-counter and requires a prescription, then the whole business model must be revised. For example, it should include medical doctors, which will prescribe such devices.

However, let's assume that the device can be sold over-the-counter. In this case, we can market and sell it straight to patients.

What channels can be used in this case? There are several approaches. For example, you can sell it through online platforms, like Amazon or Shopify. Note that these platforms have strict rules about medical devices. The seller must meet licensing requirements (have a valid license from a regulator), labeling, and marketing. Only Class I and certain Class II devices, which do not require a prescription, and are appropriately described and labeled, can be sold through these platforms. Examples of permitted devices in Canada:

- Class I devices: bandages, medical support stockings
- Class II devices: condoms, digital thermometers

The other approach is to sell it through brick-and-mortar stores. For example, Tytocare sells its medical examination kits through BestBuy in the United States, Canada, and Mexico. You can find sphygmomanometers, digital thermometers, bandages, stockings, condoms, and many other medical devices in drug stores or big box stores like Walmart or BestBuy.

Key Takeaways:

- B2C model is feasible only for Class I and certain Class II devices.
- If you consider the B2C model or consider it in the future, make sure you justify and apply for over-the-counter use in the premarket submission.

8.3   B2B Model

A much more common situation in MedDev is when a product is sold (directly or indirectly) to a healthcare organization, and ultimate users are healthcare professionals. The healthcare organization can be as small as a private practice or as big as an extensive healthcare system, encompassing multiple hospitals and other care providers. All these types of transactions fall under a broad definition of business-to-business transactions. It can also include government entities, like Veteran Health Administration (VHA) hospitals in the United States.

In order to make your device compelling for use in healthcare, it needs to be able to affect the bottom line of the business. While doctors and nurses do care about patients and better clinical outcomes, the business model built just around these items will not be convincing to investors.

Procurement in healthcare is a quite complicated process. Any purchase in healthcare is being evaluated from multiple angles. In addition to the clinical input, it will be assessed from a financial perspective, technology, etc. It holds true even if it is all done by the same person in the case of a small practice. And financial considerations, in many cases, will be the most important. Given the limited amount of resources (which is typically the case), the customer needs to decide where to allocate them. Even if there are no direct competitors or substitute technologies, in most cases, your device competes with other customer's priorities. Not to mention that in many cases maintaining the status quo can be an overarching customer's priority.

Thus, to be compelling to a customer, your medical device needs to improve its bottom line. In order to increase the profit, you can either increase revenue or decrease expenses. Therefore, your device can be compelling to a customer if it either increases their revenue or decreases expenses.

If we need to rank them, then increasing revenue is by far the most convincing argument. The reason is pretty simple. In the private healthcare environment, revenue (especially revenue growth) is an important metric used by investors and financial institutions.

Return on investment (ROI) is the second important aspect here. You need to build a compelling business case for the customer. If they spend some money today and recoup them in 20 years, it is not a compelling business case. Many things may happen during these 20 years. For example, the useful life of your product can be much shorter. Or the technology may become obsolete and needs to be replaced.

In order to build this business case for the customer, you need to understand how the customer will use your technology and how they get paid.

8.3.1 Out-of-Pocket Payments

One way how a healthcare organization can get paid is through out-of-pocket patients' payments. However, for most developed countries, it accounts only for a small fraction of payments. According to the World Health Organization Global Health Expenditure database,[12] the share of out-of-pocket expenses as a percentage of total healthcare expenditure in high-income countries was 20.5% in 2018, with a downward trend (it was 23.3% in 2000.) However, this share is typically quite large in other nations. It can be as high as 42.1% in Mexico or 62.7% in India. The statistics for several big markets are depicted in Table 8.1.

Table 8.1   Share of out-of-pocket expenditure on healthcare, 2018[13]

Country	Out-of-Pocket Share
United States	10.8%
Canada	14.7%
United Kingdom	16.7%
Germany	12.6%
Australia	17.7%
Japan	12.7%
China	35.1%
India	62.7%
Russia	38.3%
Brazil	27.5%

[12]World Health Organization Global Health Expenditure database. Retrieved from http://apps.who.int/nha/database
[13]World Health Organization Global Health Expenditure database. Retrieved from http://apps.who.int/nha/database

It should be noted that in some cases, these numbers are skewed toward particular services. For example, in Canada, most medical services are covered by government insurance and do not require co-pay. However, dental and vision services and prescription drugs are not covered by government insurance. They are either covered by private insurances, which typically require co-pay, or not covered at all, which require 100% out-of-pocket payment. Thus, Canadian data is most likely skewed toward these expenses.

8.3.2 Payers

Another common scenario is when a healthcare organization gets paid for its services through insurance payment. The organization, which pays to the healthcare organization is often referred to as an insurer or payer (or payor), and the process is often referred to as a reimbursement.

Almost all developed countries have some form of universal healthcare. That is why we will start our discussion with it.

8.3.2.1 Universal healthcare

Universal healthcare is a healthcare system in which all residents of a particular country or region are assured access to healthcare. Universal healthcare does not necessarily imply full medical coverage. Instead, it rather refers to some basic coverage.

Universal healthcare draws its history from the Sickness Insurance Law, which was introduced in Germany in 1883 by Otto von Bismarck. According to footnote 14, "industrial employers were mandated to provide injury and illness insurance for their low-wage workers. The system was funded and administered by employees and employers through "sick funds," drawn from deductions in workers' wages and employers' contributions."

Universal healthcare in most countries has been achieved by a mixed model of funding. In this case, general taxation revenue is the primary source of financing. Still, in many countries, it is supplemented by specific charges (which may be charged to the individual or an employer) or with the option of private payments

---

[14]Universal healthcare – Wikipedia. Retrieved from https://en.wikipedia.org/wiki/Universal_health_care

(by direct or optional insurance) for services beyond those covered by the public system. For example, almost all European systems are financed through a mix of public and private contributions.

A single-payer system (e.g., Canada and the United Kingdom) is one of the potential implementations of universal healthcare. Single-payer healthcare is a system in which the government, rather than private insurers, pays for all healthcare costs.

Another option is to provide universal healthcare through a choice of multiple public and private funds providing a standard service (e.g., Germany).

With the United States as a notable exception, all developed countries have some form of universal healthcare.

All UN Member States have agreed to try to achieve universal health coverage (UHC) by 2030 as part of the Sustainable Development Goals.[15]

The two primary universal healthcare models are the Bismarck model and the Beveridge model.

Bismarck Model

The Bismarck model (also referred to as the "Social Health Insurance Model") is a limited healthcare system in which people pay a fee to a fund that pays healthcare services. Services can be provided by state-owned institutions, other government body-owned institutions, or private institutions. Countries like Germany, France, Austria, Switzerland, Japan, and the Czech Republic have a Bismarck healthcare system.

Beveridge Model

The Beveridge model is a healthcare system in which the government provides healthcare for all its citizens through income tax payments. This system is implemented in Italy, Spain, Denmark, Sweden, Norway, and New Zealand. Under this system, most hospitals and clinics are owned by the government; some doctors and healthcare professionals are government employees, but there are also private institutions that collect their fees from the government.

---

[15]Universal Health Coverage, WHO. Retrieved from https://www.who.int/en/news-room/fact-sheets/detail/universal-health-coverage-(uhc)

Most universal healthcare systems (like the United Kingdom and Canada) have elements of both approaches.

Two core elements of virtually any healthcare system are public insurance and private insurance.

8.3.2.2 Public insurance and compulsory private insurance

Public health insurance is the primary method of healthcare coverage in most OECD countries. However, implementations can be very different:

- Insurance can be paid either through taxes (the United Kingdom, Canada), or people pay a fee to a fund that in turn pays healthcare services (Germany, France).
- The healthcare system may have multiple public and private funds providing a standard service (Germany) or just a single public fund (per province in Canada or per country in the United Kingdom).
- Governments can own healthcare providers (the United Kingdom) or pay private providers for services (Canada).

8.3.2.3 Voluntary insurance

In Fig. 8.1, one can see the composition of sources of healthcare funding in some OECD countries (compiled from World Health Organization Global Health Expenditure database).[16]

In most OECD countries, voluntary insurance plays a secondary role (see Fig. 8.1). Governments often look to private health insurance as a possible means of addressing some health system challenges. It plays various roles, ranging from primary coverage for particular population groups to a supporting role for public systems.

The voluntary private insurance may be complementary (meeting costs not covered by the public system such as the cost of prescription medicines, dental treatments, optometry, and co-payments) or supplementary (adding more choice of providers or providing faster access to care).

Now let us consider some country-specific examples.

---

[16]World Health Organization Global Health Expenditure database. Retrieved from http://apps.who.int/nha/database

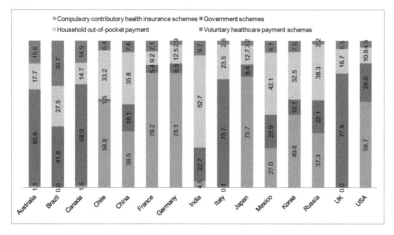

Figure 8.1 Health expenditure by source of health financing, 2018. *Data source*: WHO.[17]

United States

The United States has probably the most complicated healthcare system in the world. The core of the system is private insurance. As of 2019, there were 952 health insurance companies in the United States.[18] According to the United States Census Bureau, some 55.1% of Americans are covered through an employer, while about 10.8% directly purchase health insurance.[19]

However, multiple strata of society are covered by publicly funded insurances. As of 2018, 34 percent of Americans received their healthcare via government insurance or direct public provision.[20] According to footnote 21, "public programs provide the primary source of coverage for most seniors and low-income children and families who meet certain eligibility requirements. The primary

[17]World Health Organization Global Health Expenditure database. Retrieved from http://apps.who.int/nha/database

[18]Facts + Statistics: Industry overview, Insurance Information Institute. Retrieved from https://www.iii.org/fact-statistic/facts-statistics-industry-overview

[19]Berchick, ED, Barnett, JC, Upton, RD (2019) Health Insurance in the United States: 2018 - Visualizations. Retrieved from https://www.census.gov/library/visualizations/2019/demo/p60-267.html

[20]Berchick, ED, Barnett, JC, Upton, RD (2019) Health Insurance in the United States: 2018 - Visualizations. Retrieved from https://www.census.gov/library/visualizations/2019/demo/p60-267.html

[21]Health insurance in the United States – Wikipedia. Retrieved from https://en.wikipedia.org/wiki/Health_insurance_in_the_United_States

public programs are Medicare, a federal social insurance program for seniors (generally persons aged 65 and over) and certain disabled individuals; Medicaid, funded jointly by the federal government and states but administered at the state level, which covers certain very low-income children and their families; and CHIP, also a federal-state partnership that serves certain children and families who do not qualify for Medicaid but who cannot afford private coverage. Other public programs include military health benefits provided through TRICARE and the Veterans Health Administration and benefits provided through the Indian Health Service."

Approximately 9% of the United States population does not have any health insurance.

Canada

The backbone of healthcare in Canada is publicly funded healthcare, informally called Medicare, delivered through the provincial and territorial systems. It is universal and guided by the provisions of the Canada Health Act of 1984.

Medicare is a single-payer system in which "medically necessary" hospital and physician services are provided with fees paid for by the government. The Canada Health Act covers most healthcare (emergency services, primary care, acute care, surgeries, etc.) with a few exceptions, including prescription drugs, home care or long-term care, dental care, optometry, and physiotherapy.

About 70% of Canada's healthcare funding is via the public system. Another 30% comes from private funding, divided approximately equally between out-of-pocket funding and private insurance.[22]

Most Canadian provinces (including Ontario) banned private insurance for publicly insured services to inhibit queue jumping and preserve fairness in the healthcare system. Thus, the private insurance is typically limited to complementary services (meeting costs not covered by the public plan, such as prescription medicines, dental treatments, optometry, and co-payments) and minor supplementary services (choice of a private room in the hospital).

Approximately 65 to 75 percent of Canadians have some form of complementary or supplementary health insurance; many receive it

---

[22]Exploring the 70/30 Split: How Canada's Health Care System Is Financed, (2005) Canadian Institute for Health Information. Retrieved from https://secure.cihi.ca/ free_products/FundRep_EN.pdf

through their employers or use secondary social service programs related to extended coverage for families receiving social assistance or vulnerable demographics, such as seniors, minors, and those with disabilities.[23]

Healthcare providers: in Canada, there are private and public healthcare providers. Patients have complete freedom of choice between doctors and facilities.

70% of MedTech purchases in Canada are made at the hospital level. Hospitals receive an annual global operating budget as a lump sum from the government based on factors such as:

- Past allocations
- Characteristics and size of population served
- Priorities set forth by the hospital
- The expertise of medical professionals in the hospital

United Kingdom

Healthcare in the United Kingdom consists of publicly funded healthcare, a smaller private sector, and voluntary (charity) provisions.

The National Health Service (NHS) is the umbrella term for the publicly funded healthcare systems of the United Kingdom. Each country of the United Kingdom (England, Northern Ireland, Scotland, and Wales) has its own publicly funded healthcare systems, funded by and accountable to separate governments and parliaments.

NHS is the first universal state-funded healthcare system (Beveridge model). It was established in 1948.

According to some industry estimates, the United Kingdom private healthcare market had been forecast to grow from $11.8bn in 2017 to $13.8bn by the end of 2023.[24]

Examples of the benefits provided by complementary and supplementary voluntary health insurance in the United Kingdom are depicted in Table 8.2.

---

[23]Tsai-Jyh Chen (2018). An International Comparison of Financial Consumer Protection. Springer. p. 93.

[24]"UK Private Healthcare Market Expected to Reach US$ 13.8 Bn by End of 2023 with CAGR of 2.6%." Medgadget. Retrieved from https://www.medgadget.com/2018/07/u-k-private-healthcare-market-to-reach-us13-8-bn-by-2023-due-to-rising-nhs-waiting-lists-top-players-are-aspen-healthcare-bmi-healthcare-care-uk-and-other.html

Table 8.2 Examples of the benefits provided by complementary and supplementary voluntary health insurance in the European Union and the United Kingdom. *Source:* Adapted from footnote 25. Copyright WHO

Country	Complementary	Supplementary
Austria	- hospital per diem charge (cash benefit) - alternative treatment	- physician costs - supplementary hospital costs - faster access/increased choice
Belgium	- legal co-payments for non-reimbursed in/outpatient costs - carer costs (loss of independence)	- supplementary hospital costs
Denmark	- co-payments for drugs, dental care, physiotherapy, spectacles, etc.	- access to private hospitals in Denmark and abroad
Finland	- some public-sector hospital costs - travel expenses	- private care for children - faster access - increased choice (including access to private hospitals)
France	- co-payments (including differences between negotiated and real prices) - treatments excluded by public sector - home help - hospital per diem charge	- faster access to specialist consultations - choice of a private room in a hospital

[25]Mossialos, E. and Thomson, S. (2004). Voluntary health insurance in the European Union. World Health Organization. Regional Office for Europe. Retrieved from https://apps.who.int/iris/handle/10665/107601

(Continued)

Table 8.2 *(Continued)*

Country	Complementary	Supplementary
Germany	- outpatient care - dental care - hospital daily allowance (cash benefit)	- choice of specialist - amenity beds
Greece	- hospital daily allowance (cash benefit)	- faster access - choice of private provider and accommodation
Ireland	- outpatient cover for GP visits, specialist consultations, X-rays, and other items - outpatient cover for alternative treatment (BUPA Ireland) - hospital per diem charge - cost of OT fees, X-rays, lab tests, drugs in a hospital - consultants' fees for inpatient, daycare, and some outpatient treatment - maternity benefits - convalescence in a nursing home	- cost of hospital accommodation in private beds in public hospitals and private hospitals (including daycare surgery)
Italy	- co-payments - non-reimbursed services - dental care - hospital per diem charge	- increased choice of provider - increased access to private hospitals

Country	Complementary	Supplementary
Luxembourg	- hospital co-payments - pre-and post-operative and convalescence costs - dental prostheses - surgical treatment abroad - partial reimbursement where no agreement on the cost of a treatment	- additional charges for a private room in a hospital
Netherlands	- mainly dental care - drug co-payments (marginal) - cross-border care - alternative treatment	- faster access to acute and long-term care
Portugal	- dental care - ophthalmology - co-payments - cash benefits	- access to private providers
Spain	- dental care	- increased choice of provider
Sweden	- some reimbursement of co-payments, drugs, dental care, alternative treatment	- faster access to elective outpatient care - access to private hospitals
United Kingdom	- cash benefits - dental care - alternative treatment	- faster access to specialists and elective treatment - choice of amenities in public hospitals

European Union

Healthcare in Europe is provided through a wide range of different systems run at individual national levels. Most European countries have a system of tightly regulated, competing private health insurance companies, with government subsidies available for citizens who cannot afford coverage.[26]

Examples of the benefits provided by complementary and supplementary voluntary health insurance (VHI) in the European Union are depicted in Table 8.2.

8.3.3 Reimbursement

In the previous discussion, we figured out that health insurance (either public or private) plays a dominant role in virtually any OECD country, except for Mexico, where it is on par with out-of-pocket payments. Thus, it is pretty important to understand how the insurance model works in healthcare. It will be our focus for the rest of this chapter. We will do it using the United States as an example.

In a typical scenario, a healthcare organization provides healthcare services first. Then, the provider submits a claim to an insurer. The insurer processes the claim and, if it finds the claim eligible, pays to the provider. As we mentioned earlier, the insurer, which pays to the healthcare organization, is often referred to as a payer, and the process is often referred to as a reimbursement.

As we mentioned, in 2019, there were 952 health insurance companies in the United States. If we have that number of entirely independent players and each comes with its own model, it will be an insanely complicated market. Not to mention the challenge for any MedTech entrepreneur: where to start? However, it does not happen. While the insurance market is very complex and insurance coverage varies between payers, these variations are minor. How do they coordinate their insurance coverage?

While politicians in the United States like to emphasize that private insurance is at the core of their system, nevertheless, the beacon of this system is a government agency, Centers for Medicare

---

[26]Sanger-Katz, M. (2019). "What's the Difference Between a 'Public Option' and 'Medicare for All'? Retrieved from https://www.healthline.com/health/medicare/medicare-for-all-vs-public-option

& Medicaid Services or CMS, which manages the Medicare and Medicaid programs.

Thus, any MedTech company's typical goal is to get their product/service reimbursable by Medicare. After that, the product/service will be gradually pulled into the system and covered by commercial payers.

Reimbursement consists of three interrelated but not consequential elements: coverage, coding, and payment. Let's consider them in some detail.

8.3.3.1   Coding

Medical codes are used on claims for reimbursement to identify the procedure and diagnosis for the underlying condition.

There are several types of codes, which are used for different purposes and in different care settings:

Diagnostics Codes: ICD-10

- ICD stands for International Classification of Diseases
- These codes describe the diagnosis and patient condition
- They are oversighted by World Health Organization and Center for Disease Control (in the United States)

Procedure Codes: CPT ICD-10

- CPT stands for Current Procedural Terminology.
- These 5 digit codes describe the procedure or service.
- American Medical Association (AMA) controls the issuance of CPT codes. An example of CPT code:
  27130–Arthroplasty, femoral prosthetic replacement ("total hip")

Product-specific codes HCPCS

- HCPCS stands for Healthcare Common Procedure Coding System
- They describe the actual device or technology type used
- CMS controls HCPCS codes

HCPCS codes refer to a broader set of codes, which is used as transactional codes in reporting. They include CPT as Level I codes. Level II codes "cover non-physician products, supplies, and procedures not included in CPT. Level III codes, also called HCPCS local codes, were developed by state Medicaid agencies, Medicare contractors, and private insurers for use in specific programs and jurisdictions."[27]

For novel technologies, CMS assigns Q codes. Q codes are assigned to procedures, services, and supplies on a temporary basis. "The Q codes are used to identify services that would not be given a CPT code, such as drugs, biologicals, and other types of medical equipment or services, and which are not identified by national Level II codes. However, these services and equipment need codes for claims processing purposes."[28] When a permanent code is assigned, the Q code is deleted and cross-referenced.

Different care settings use different codes. For example, hospitals use ICD-10-CM diagnosis and ICD-10-PCS procedure codes in inpatient settings. However, physician offices use CPT codes.

Procedure codes and product-specific codes have the primary relevance to MedTech entrepreneurs. For example, CPT codes are created and maintained by a working group of the American Medical Association, AMA. It can be a significant task to convince them to make a new code. It also may take a substantial amount of time. Thus, using existing codes, if possible, simplifies things significantly.

8.3.3.2 Payment

Payment refers to the actual amount paid for the procedure. The practical question here is a delicate balance between physicians and hospitals are paid enough to encourage product adoption, or it is too expensive and discourages government and private insurance coverage.

Payment and payment methodology depend significantly on the care settings. In particular, there are various payment methodologies for:

- Inpatient Hospital Services

---

[27]Note similarities and differences between HCPCS, CPT … . Retrieved from https://www.hcpro.com/HIM-284009-8160/Note-similarities-and-differences-between-HCPCS-CPT-codes.html

[28]Note similarities and differences between HCPCS, CPT … . Retrieved from https://www.hcpro.com/HIM-284009-8160/Note-similarities-and-differences-between-HCPCS-CPT-codes.html

- Outpatient Hospital Services
- Ambulatory Surgery Centers (ASC)
- Physicians
- Durable Medical Equipment, Prosthetics, Orthotics and Supplies (DMEPOS)

Hospital Inpatient Payment Methodology

The key features of Hospital Inpatient Payment Methodology are:

- "Payment system: Inpatient Prospective Payment System (IPPS).
- Method: Medical Severity-Diagnosis Related Groups, MS-DRGs are the predominant method for paying for inpatient hospital services.
- Coding: ICD-10-CM diagnosis and ICD-10-PCS procedure codes."[29]
- "There are around 900 DRGs to which any medical or surgical admission will be assigned based on patient diagnoses, procedures. Each DRG has the predetermined resources it should require to take care of the average patient.
- Each DRG has a "weight" that is multiplied by a conversion factor to determine payment.
- New Tech Pass-Through Payment allows the cost of new tech to be paid at/near the retail price for a period of two years if it meets a cost threshold and achieves "substantial clinical improvement.
- As a result, hospitals will adopt less expensive technologies wherever they can because their revenue is fixed." [30]

Hospital Outpatient and ASC Payment Methodology

The key features of Payment Methodology for Hospital Outpatient and Ambulatory Surgery Centers are:

- "Payment system: Outpatient Prospective Payment System (OPPS)
- Method: APCs –Ambulatory Payment Classifications

---

[29]MCRA, Evidence generation for optimal market adoption, LSX London 2021 Medtech, Feb 18, 2021.
[30]Black, E., Reimbursement Strategies, MedTech/BioTech Reimbursement: Getting Paid in the USA, Sept 2016.

- Coding: CPT, HCPCS, and ICD-10-CM diagnosis codes" [31]
- "APCs work similarly to DRGs (inpatient services). There are about 350 APCs that, similarly to DRGs, are undifferentiated for Complications/Comorbidities
- Medicare processes all outpatient hospital claims using APCs; private payers often adopt their own hybrid methodologies
- Hospitals bill CPT codes, which get mapped to one of the 350 APCs. Consequently, payment for therapeutic and diagnostic services are not highly differentiated
- A New Technology Add-On Payment is available for new technological devices that demonstrate "substantial clinical improvement" and whose costs are "not insignificant" to the APC payment. However, these are high thresholds to achieve, and the Add-On Payment expires after 2–3 years." [32]

Physician Office

- Payment system: Physician Fee Schedule (PFS)
- Method: Relative Value Unit (RVU) assigned to CPT codes
- Coding: CPT
- The payment consists of RVU multiplied by a conversion factor
- RVUs are consistent among government and private payers
- Conversion Factors, consequently allowances, vary by payer

The differences in payment methodologies lead to entirely different financial outcomes in different care settings. It was found [33] that the payment differences between various outpatient settings were significant. Medicare reimbursement for similar physician services would have been $114 000 higher per physician per year if a physician were integrated (the practice belongs to a hospital system) compared to being non-integrated (independent practice). Primary care physicians faced a 78% increase, medical specialists 74%, and surgeons 224%.

---

[31]MCRA, Evidence generation for optimal market adoption, LSX London 2021 Medtech, Feb 18, 2021.

[32]Black, E. Reimbursement Strategies, MedTech/BioTech Reimbursement: Getting Paid in the USA, Sept 2016.

[33]Post, B., Norton, E. C., Hollenbeck, B., Buchmueller, T., Ryan, A. M. Hospital-physician integration and Medicare's site-based outpatient payments. *Health Serv Res.*, 2021, 56: 7–15.

Key Takeaways:

- The Fee for Service (FFS) is the primary payment system for physicians. DRG/APC systems are primary for hospital settings.
- The importance of the setting of care is closely related to coding and payment. Payment levels and restrictions on eligibility may be structured so that it is not economically feasible for a technology's intended setting of care.
- In some cases, early adoption is easier in a different setting of care due to the existence of applicable procedure codes or the healthcare system's culture.

8.3.3.3 Coverage

Coverage refers to whether Medicare and private insurers cover the procedures associated with the devices and, if so, under what clinical circumstances. Not all services that have a CPT code and payment allowance are covered.

There are several empirical facts about coverage in the United States:

- Medicare is the largest single-payer in the United States.
- Private insurers often, but not always, follow Medicare coverage decisions.
- Government payers (Medicare, Medicaid, TRICARE, and VA) all pay less than private insurers for the same services.

Medicare has two primary mechanisms to establish coverage: National Coverage Determination (NCD) through CMS and Local Coverage Determination (LCD) through Medicare Administrative Contractors (MAC). MACs also can perform case-by-case adjudication.

According footnote 34, for National Coverage Determination (NCD):

- "Coverage is granted to procedures and technologies when they meet the definition of "reasonable and necessary"
- Often under Coverage with Evidence Development (CED) protocol
- Reserved for high profile, high impact, controversial technology

---

[34]MCRA, Evidence generation for optimal market adoption, LSX London 2021 Medtech, Feb 18, 2021.

- Medicare National Determinations are binding for all MACs
- Typically 9–12 months to establish"

Local Coverage Determination (LCD) is an alternative mechanism granted by Medicare Administrative Contractors (MAC). The United States is split into 10 jurisdictions administered by eight private contractors—all with their own medical policy staffs. These 10 jurisdictions are depicted in Fig. 8.2.

According to footnote 35, for LCD:

- "MACs established LCDs for selective services with criteria outlining under what conditions the technology is "reasonable and necessary."
- MAC may or may not have formally published LCDs.
- Inconsistent coverage—LCDs can vary across different jurisdictions."

The evidence requirements (which will be discussed next) differ between different MACs. For example, some MACs require peer-reviewed literature, while others require white papers.

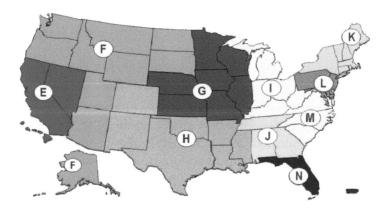

Figure 8.2 Consolidated Medicare A/B MAC jurisdictions. Reproduced from footnote 36.

---

[35]MCRA, Evidence generation for optimal market adoption, LSX London 2021 Medtech, Feb 18, 2021.
[36]Retrieved from https://www.cms.gov/Medicare/Medicare-Contracting/Medicare-Administrative-Contractors/Downloads/Consolidated-AB-Map-Vision.pdf

8.3.3.4 Regulatory approval vs. reimbursement approval

Reimbursement approval does not follow regulatory approval automatically. These are two distinct activities, which are decoupled to some extent. They are being performed by entirely different bodies, which follow a completely different set of rules.

For example, in the United States, regulatory approval is being performed by the FDA. The FDA is a science-driven organization that cares that the product is "safe and effective." The "reimbursement approval" is being performed by CMS. CMS cares that the application of the device is "reasonable and necessary."

As you may see, the FDA and CMS have entirely different sets of objectives.

Like CMS, US private payers are interested in health economics and clinical outcomes of the proposed technology. For example, below are typical commercial payer technology evaluation criteria:[37]

- The technology must have final approval from the appropriate regulatory body
- The scientific evidence must permit conclusions concerning the effect of the technology on health outcomes
- The technology must improve the net health outcome
- The technology must be as beneficial as any established alternatives
- The improvement must be attainable outside the investigative setting

While the regulatory approval addresses the first item in this list, the remaining four criteria may require extra work. For example, the "net health outcome" will necessitate comparison with the existing standard of care. Thus, in the optimal scenario, it will require a randomized clinical trial. Moreover, the improvement should be statistically significant, which may pose additional requirements on the number of subjects.

Various payers have slightly different requirements for the data quality for coverage. In general, good quality of evidence (a well-designed study and peer-reviewed publications) is required. However, some payers, like CMS, can be approached initially at

---

[37]How we evaluate new technology, Wellmark. Retrieved from https://www.wellmark.com/evaluating-technology

earlier stages of the process (data gathering). Some large private payers even have research arms, which can help with designing clinical studies. However, in the case of smaller payers, it is better to approach them with already published or at least accepted for publication results. So, it is recommended to do homework before approaching any payer.

As we said already, payers are looking into clinical outcomes and health economics. Clinical benefits are primary and economic benefits are secondary. But it should be a clinically meaningful difference (from the statistical point of view).

How much data are required? In general, payers use the following approach. If a procedure decreases the overall cost of patient treatment, then data requirements will be less stringent. However, if the procedure or the device increases overall costs, then payers will require more durable data to support the net health outcome.

The US is not unique here in any way. To get the product use reimbursed in other jurisdictions, it is typically required to perform some sort of "Technology Assessment." For example, it is called a Health Technology Assessment or HTA in Canada. HTA is a method of evidence synthesis that considers evidence regarding clinical effectiveness, safety, cost-effectiveness and, when broadly applied, includes social, ethical, and legal aspects of the use of health technologies. The primary focus of HTA is on cost-effectiveness/cost-utility. HTA processes in Canada are highly decentralized. HTAs are conducted at national, provincial, and hospital levels. A major use of HTAs is in informing reimbursement and coverage decisions. Decisions to participate or adhere to recommendations from province-wide review by hospitals and administrators may still be voluntary depending on the province.

A similar centralized process exists in the United Kingdom. The National Institute for Health and Care Excellence (NICE) was established in 1999 to guide the NHS on technologies' clinical and cost-effectiveness. NICE routinely evaluates new technologies and treatments and makes recommendations to the NHS. NICE conducts five types of reviews:

- Medical Technology Evaluation Programme (MTEP);
- Clinical Guidance (CG);
- Technology Appraisals (TA);

- Interventional Procedures Guidance (IPG); and
- The new Diagnostic Assessment Programme (DAP).

NICE guidance is advisory except for the technology appraisal, which is mandatory in the NHS.

In addition to the health economics review, which is mandatory for many single-payer health systems, there is a clinical guidance review in many countries. While it is not always required, it still can impact technology adoption significantly.

The focus of the clinical guidance review is to compare the proposed device with existing alternatives. At least, it should demonstrate non-inferiority. This type of review focuses on peer-reviewed publications with randomized clinical trials only. For example, the NICE Clinical Guideline group routinely declines to review any single-arm studies.

Key Takeaways:

- Reimbursement approval is entirely different from regulatory approval and may require completely different evidence.
- Most entrepreneurs take a sequential approach to market access planning and fundraising, tackling regulatory approval first and then turning their focus to reimbursement. This approach is suboptimal and may have severe time and cost implications.
- You need to start thinking about how to generate evidence for the reimbursement approval beforehand.

Chapter 9

Other Considerations

In previous chapters, we have considered several factors, which are critical for MedTech startup's success. Now, it is a turn to do incursions into several other less obvious but still very important factors.

There are multiple studies about the reasons for startup successes and failures. In one such study,[1] Bill Gross from Idealab analyzed 200 startups (both successful and failed) and identified five primary reasons for startup's success: business model, funding (or lack of thereof), team, idea, and timing. These five drivers and their relative importance are depicted in Fig. 9.1.

As you may notice from Fig. 9.1, funding and business model, which we discussed in Chapter 6 (Funding) and Chapter 9 (Business Model), respectively, do not play the primary role in the startup's success. It is quite an astonishing finding. The primary culprit here is the subset of startups, which was analyzed. Most of the names researched by Gross belong to business and consumer services and do not include any notable MedTech name. I would argue that due to the very long road to the market (8–10 years as we found in Chapter 5 (Timelines and Capital)), the business model and funding roles are much more profound in the MedTech startups. However, the three remaining factors (team/execution, idea, and timing) are still very important.

---

[1]Bill Gross, TED Talk, March 2015, Vancouver, BC, Canada.

Bringing a Medical Device to the Market: A Scientist's Perspective
Gennadi Saiko
Copyright © 2022 Jenny Stanford Publishing Pte. Ltd.
ISBN 978-981-4968-25-6 (Hardcover), 978-1-003-31221-5 (eBook)
www.jennystanford.com

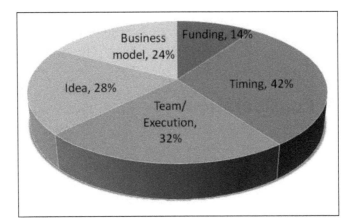

Figure 9.1 Primary reasons for startup's success: business model, funding, team, degree of innovation, and timing.

This chapter discusses these three factors (team/execution, idea, and timing) and briefly touches on several other aspects relevant for the MedTech startups.

9.1 Team

Bringing a MedDev product to the market is a long and enduring process. It will be challenging to undergo it alone. It is much more fun to do it in good company. That is why the team is vital. There are several critical aspects here.

9.1.1 Skin in the Game

Investors like to see the skin in the game from founders. Some of them (particularly at very early stages, like angels) invest their own money, so they want to see that the founding team is committed to the startup. That is why in many cases, they will not be impressed by a busy university professor or practicing surgeon, who also serves as a CEO to the startup. They would definitely prefer a full-time CEO who is committed to the company's success.

It is a true statement in general. If you expect that you will be able to do a startup in your spare time at night and on weekends, it is a big mistake. A MedTech startup requires a lot of commitment. So, unless you have full-time people committed to its success, it is rather a hobby. And make no mistake, it will be perceived as a hobby by external parties.[2]

9.1.2   Knowledge Transfer

Knowledge accumulation and knowledge transfer are other vital areas. As we mentioned in Chapter 7 (IP and Other Moats), the special knowledge and know-how accumulated in the company can be a significant competitive advantage and moat, which can deter others from entering the market. It is much easier to achieve with the help of the permanent team.

The use of contractors and temporary staff is completely justifiable for many startups. With the lack of stable funding, oftentimes, it is the only way to go. However, this model has several apparent downsides. Firstly, efficient knowledge transfer is quite problematic in such a model. With incoming and outgoing contractors, definitely not all knowledge accumulated is passed from one individual to another. The code and some documentation may stay in the repository; however, multiple little details will be lost in any transition. Secondly, it leads to a longer learning curve for any new employee or contractor. Without support from an experienced team member, the sifting and troubleshooting of somebody else code can be painful, especially in the absence of good documentation. Thus, the productivity of a new employee or contractor can be significantly impaired at the beginning. Two just mentioned factors lead us to the third drawback, the knowledge accumulation happens at a completely different pace, much slower. Finally, keeping the intact intellectual property in general and trade secrets, in particular, is much more challenging with constantly changing employees and contractors.

---

[2]There are known cases where founders were part time and became successful. According to Forbes, Warby Parker (sunglasses) is such an example, where none of the four co-founders pursued the opportunity full-time. This example came to me from an investor who passed on them because of that, and then they turned very successful. But it is exception to the rule, and that was why the investor was talking about it.

9.1.3   Tacit Knowledge

The presence of the permanent team has more benefits than efficient knowledge transfer. Creating a team and culture within the organization may have a very beneficial additional by-product: tacit knowledge. Tacit knowledge refers to something that cannot be easily communicated, either verbally or in writing. As pointed out,[3] tacit knowledge can present a significant competitive advantage to the company. For example, in a company, tacit knowledge can include manufacturing techniques, testing procedures, customer insight, and supplier dynamics. In general, any field where "there is more art than science" is probably one where tacit knowledge is important.

While tacit knowledge can be accumulated on the individual level, the collective tacit knowledge is more robust as a moat. This sort of knowledge cannot be obtained by competitors through industrial espionage and not always by hiring away employees.

Typically, tacit knowledge cannot be built immediately. However, with time it may become an important competitive advantage.

9.1.4   Balanced Team

Oftentimes, startup teams resemble a dental flux. They are not symmetrical. Oftentimes, they have most or all co-founders are coming from the same field and similar skill sets. For example, they come from the same class or the same lab.

In many cases, such team composition is suboptimal. While it is easier to get an agreement between people, who share the same background, it also can be much easier to overlook certain problem if it arises.

A more optimal situation is to have several members of the founding team with diverse backgrounds. The diversity allows you to have more multi-facet vision and avoid certain blind spots. Below we discuss several points to consider.

In technology-driven startups, it is always helpful to have a co-founder or member of the founding team with IT (particularly coding) skills. This person is always around and can rapidly

---

[3]http://reactionwheel.net/2019/09/a-taxonomy-of-moats.html

prototype a solution or fix a problem with an existing code. And he/she can do it even if the company is squeezed and does not have any money to pay contractors or salary to employees.[4]

Having somebody with a medical background on the team is of importance for any MedTech startup. The healthcare world is very complicated, and having somebody who understands these complexities and can be a liaison with this world can be of great importance for success. It is also important for investors. Thus, if you do not have a co-founder or member of the founding team with a relevant background, consider engaging a medical advisor.

An experienced entrepreneur is another helpful addition to the team. It can appease investors and increase the investability of the company. If you do not have a co-founder or member of the founding team with a relevant background, consider engaging them as mentors or advisors.

Another important consideration is to have a multi-gender and multi-ethnicity team. It is particularly important nowadays with the consistent push for gender-balanced teams. However, it is not just about optics; it is your competitive advantage. People with various backgrounds have different angles to look at a problem, understanding different niche markets and lessons from them. These niche markets may become your beachhead markets. They also have different personal and professional networks. Thus, you immediately broaden your network significantly. This can put advisors, investors, suppliers, contractors, etc., within reach immediately.

9.1.5 Believers

The team and particularly the early team should be composed of true believers in the company and its technology.

MedTech entrepreneurship is a long and occasionally bumpy journey. So, it will be mentally challenging, painful, and detrimental to the team's morale to constantly overcome internal resistance and skepticism. It is much better if you can rely on the support from every team member and they can support each other.

---

[4]I know personally several technology companies where CEOs write the code, at least occasionally.

9.1.6 Setting Expectations

As we mentioned multiple times already, MedTech startup can be a very long venture. To avoid issues in the future, you have to set expectations within the team, particularly among founders. If you have the vision to build a company, which will revolutionize healthcare, it may take 10+ years. It can be quite problematic if somebody on your team (e.g., your hired CEO) expects to make a quick buck within three years.

So, expectations should be discussed with all critical stakeholders, e.g., founders, investors, board members, critical employees.

9.1.7 Aligning Incentives

Aligning incentives is closely related to setting expectations within the team. The MedTech may take many years, so that personal circumstances may change. The last thing you want is to have idle co-founders who lost interest in the business but still own a significant part of the company.

One way to address it is to align incentives. If you want people to contribute to the company in the long term, it is better to put a structure. For example, for co-founders or founding teams, you can vest founders' shares[5]. For instance, stocks can be vested monthly for four years. Another option is to be able to vote out a co-founder. For example, if you have more than two co-founders, the majority of co-founders may vote out a non-performing founder.

All these conversations need to take place at the beginning of the venture and should be documented. Do not expect that somebody will give up their shares in the company voluntarily. It will be a heated discussion in most cases, so it is better to put the written agreements in place beforehand.

9.1.8 Solo Founders

Occasionally, you may end up as a solo founder of the company, as it was in my case. This situation has several advantages.

---

[5]Share vesting is the process by which an employee, investor, or co-founder is rewarded with shares or stock options but receives the full rights to them over a set period of time or, in some cases, after a specific milestone is hit. (What Is Share Vesting? Linkilaw. https://linkilawsolicitors.com/insight/what-is-share-vesting/)

The primary advantage is that the decision-making is much simpler. You can move blazingly fast.

You also have more shares in the company.

However, despite several advantages, it also has significant drawbacks.

Your opinion can completely blindside you. With no feedback from co-founders, you can overlook important aspects. Thus, simplified decision-making may become a disadvantage.

During the long startup journey being alone is mentally tricky. You will undoubtedly have multiple ups and downs. So, being alone on this roller-coaster can be quite challenging.

The solo founder is perceived as a significant risk by investors. What will happen to the company (and their money) if something happens to you?

So, in a nutshell, being a solo founder is not the most optimal solution. It increases risks significantly and decreases your investability. From the ownership perspective, owning 40% or 20% of something big is much better than owning 100% of nothing.

If you ended up in the solo founder situation, then having an independent opinion is getting even more critical. That is why you need to have advisors and/or mentors. Also, consider having a strong founding team.[6]

9.2   Degree of Innovation

When we think of innovators, we typically think about a radical idea, which has a significant impact on humankind. For example, Alexander Graham Bell, who invented the telephone, or the Wright brothers, who invented an aircraft. However, in addition to a few very radical ideas, there are many more innovations around them. For example, these innovations gradually transformed the original Wright brothers' aircraft into modern airliners.

The same logic can be applied to companies. Oftentimes, people talk about innovative and not innovative companies. To some extent, it is a misconception. Most companies are innovative to a certain degree. For example, many of them constantly update their

---

[6]A founding team member is an early employee, not a founder.

products, which in many cases represent an innovation. However, not all inventions are being created equal. They differ in an aspect, which can be called a "degree of innovativeness."

Interestingly enough, not all great inventions are patentable. Jonas Salk famously did not patent the polio vaccine and positioned himself as a benevolent inventor. However, evidence[7] shows that his lawyers investigated it and realized it was not possible to patent it due to a lack of innovation. Nevertheless, he became a famous inventor of the polio vaccine.

To categorize the degree of innovativeness, innovations are typically analyzed across two dimensions: Technology and Market. It results in a 2×2 matrix, which gives us the following four types of innovations (see Table 9.1).

Table 9.1 Degrees of innovation

	New	**Architectural Innovation**	**Disruptive Innovation**
Market	Existing	**Incremental Innovation**	**Radical Innovation**
		Existing	New
		Technology	

Incremental innovations refer to innovations, which apply improved existing technology to the existing market. They demonstrate a minimal degree of innovation. However, they represent the bulk majority of all innovations. Typically, these are minor changes to existing products, processes, or business models, which the company applies to a known or similar market. Usually, they do not imply new technology. Instead, it further develops known technologies, products, services, processes, or business models. Typically, they reduce costs or functional improvements in existing products and can be characterized by small uncertainty and technology risks.

A so-called "architectural innovation" stems from the application of the existing technology to a new market. Many military technologies fall into this category. Internet and GPS are classic

---

[7]The Real Reason Why Salk Refused to Patent the Polio Vaccine (2012). Retrieved from https://www.bio.org/blogs/real-reason-why-salk-refused-patent-polio-vaccine

examples of such inventions. However, the Internet later became a disruptive innovation.

Radical or breakthrough innovations refer to innovations, which apply a novel technology to the existing market. They show a very high degree of innovation (sometimes they also refer to as disruptive innovations) and represent major technology advances in the field.

Disruptive innovation creates a new market and value network and eventually disrupts an existing market and value network, displacing established market-leading firms, products, and alliances.[8] They often have their starting point in a barely recognized niche. Initially, they typically address only a small number of customers before becoming a dominant market factor and displace established companies from the market. So, in many cases, disruptive innovation can be identified in hindsight. For example, smartphones were initially considered a radical innovation. However, with their proliferation, they disrupted multiple markets.

Some examples of disruptive innovations:

- Personal computers disrupted the market of mainframes
- Semiconductor transistors disrupted the market of vacuum tubes
- In medicine, ultrasound disrupted the market of radiography (X-rays)
- In another example from medicine, coronary stenting disrupted the cardiac surgery market

Clayton M. Christensen from Harvard University introduced the idea of disruptive innovations. In his 1997 book, "The Innovator's Dilemma," he outlined his theory about the impact of disruptive innovation.

Note that not all innovations are disruptive, even if they are revolutionary. The most cited example here is automobiles. Invented in the 19th century, they were produced manually and were luxury items that cannot disrupt the horse-driven vehicles (carriage) market. However, Henry Ford disrupted this market with a mass-produced automobile (Model T).

Some authors group innovations differently. For example, architectural and radical innovations are combined in the same

---

[8]Ab Rahman, Airini; et al. (2017). "Emerging Technologies with Disruptive Effects: A Review." *PERINTIS eJournal*, 7(2).

group (oftentimes called "breakthrough innovation"). In this case, innovation can be classified as minor (incremental), moderate (breakthrough), and major (disruptive). These three types of innovations and their occurrence are depicted in Figure 9.2.

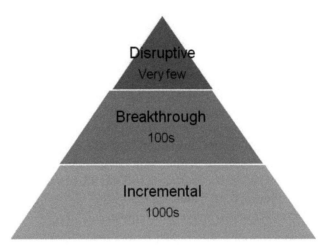

Figure 9.2   Degrees of innovation and their occurrence.

The investment potential of a startup depends significantly on the degree of innovation.

If you have a disruptive idea, most likely, you do not need to look for investors. More often, they look for you already.

If you have a breakthrough or radical idea, your investability is high. You will have a pretty high response rate from investors.

However, the vast majority of startup ideas belong to incremental ideas. In this case, you need to push hard to get investors' attention. And remember, as a rule of thumb, to bypass competition, you need to provide a so-called exponential advantage, roughly 10× improvement in something, e.g., cost or throughput.

9.3   Timing

Timing of the product is of paramount importance for the startup's success. As we mentioned earlier in this chapter, Bill Gross found that timing is a key factor in astonishing 42% of all successes and failures.

There are multiple examples of right timing among well-known companies:

YouTube was not the first video-hosting platform. Before it, there were several other products like Z.com, Mevio, Blip.tv, and Google Video (US), or Videolog (Brazil). However, YouTube's introduction in 2005 coincided with the mass availability of broadband internet, which propelled its use. According to oecd.org, from 2000 to 2005, the number of broadband connections in all 34 OECD countries grown from 20 million to130 million.

Similarly, social media were not invented by Facebook in 2004. For example, LinkedIn was launched in 2002. However, if LinkedIn is a business-oriented platform, now-defunct Myspace (launched in 2003) was in the same space as Facebook. Thus, at least some of Facebook's success can be attributed to the mass availability of broadband internet, which was further propelled by smartphone adoption (iPhone was introduced in Jan 2007).

Airbnb was launched in August 2008. It had a few stumbles at first, but in Jan 2009, founders got seed funding from Y Combinator, which helped them raise $20,000. Then, in April 2009, the company gained a $600,000 seed investment from Sequoia Capital. By 2011, the company was in 89 countries and was valued at $1 billion. Thus, this meteoric rise coincides with the aftermath of the Great Recession of 2008/2009, when the struggling homeowners were in a desperate search for additional income sources.

The same crisis helped another iconic company, Uber, to get off the ground. Uber was founded in 2009, and the mobile app was officially launched in 2011. Again, the quick wide adoption of the business model by car owners was at least in part caused by their desire to find additional sources of income.

The whole term "gig economy" can be traced back to 2009, when it became mainstream.

Another interesting example of product timing on a smaller scale is "Snuggie," a sleeved blanket. According to footnote 9, in late 2008 and early 2009, the Snuggie brand became almost a pop culture phenomenon. However, its almost identical predecessor Slanket was on the market for 10 years already, since 1998.

---

[9]Sleeved blanket – Wikipedia. Retrieved from https://en.wikipedia.org/wiki/Sleeved_blanket

The list goes on and on. The important lesson here is that timing is very important. Such as it is predominantly out of your control, you have to be very honest to yourself whether the market and customers are ready for your invention. If they are not ready, even billion dollars budget will not prevent failures.

For example, with its multi-billion dollar research budget, IBM overpromised and under-delivered with their AI-driven healthcare system Watson introduced in 2010. They have not accounted that regulators were not ready to approve AI-based systems. Now, hopefully, things have been changed, and FDA started approving AI-based medical devices.

In the non-medical world, the known failure is Google Glasses. According to footnote 10, Google started selling largely anticipated Google Glasses to qualified "Glass Explorers" in the United States on April 15, 2013, for a limited period for $1,500, before it became available to the public on May 15, 2014. The headset raised significant concerns that its use could violate existing privacy laws. As a result, on January 15, 2015, Google stopped producing the Google Glass prototype. In 2017, Google entered the corporate market with an enterprise version.

Timing may not just open an opportunity, as in the examples above. It may also close it if the public perception moved or alternative technology arrived. Such an example of a large-scale failure is Alphabet's Loon project. Google started the Loon project in 2011 to use high-altitude balloons in the stratosphere at an altitude of 18 km (11 mi) to 25 km (16 mi) to create an aerial wireless network for remote areas. The project was shut down quietly in 2021, which can be probably partially attributed to Elon Musk's Starlink project.

9.4 Therapy vs. Diagnostics

Typically, medical technologies are split into four broad categories: therapeutics, diagnostics, devices, and digital health. However, in this section, in addition to therapeutics vs. diagnostics, we will talk about medical devices' sub-categories: therapeutic and diagnostic devices.

---

[10]Google Glass – Wikipedia. Retrieved from https://en.wikipedia.org/wiki/Google_Glass

All my experience in MedDev shows me that diagnostics is a second-class citizen. The reason for that lays probably in human nature. We are ready to pay virtually any price for a magic cure.

Thus, we know multiple examples when the annual cost of treatment can be in the hundred thousand dollars price range. Moreover, in many cases, where there is no competition, the treatment price is entirely inelastic. A manufacturer can increase the price significantly with little impact on the number of units sold. This idea is not completely hypothetical and can be illustrated in the EpiPen example, a life-saving medication for children and others facing fatal allergies with little real competition. In Fig. 9.3, one can see that in 2007–2016, the price of EpiPen was raised 6.5× since low $50s to more than $350. Since the end of 2013, the price has been raised by 15 percent every other quarter.

EpiPen Prices Hiked Over 500%

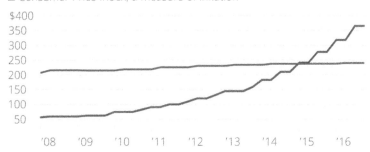

Figure 9.3   EpiPen wholesale price vs. Consumer Price Index in 2007–2016 in the United States. Reproduced from footnote 11. Copyright NBC.

In many cases, a diagnostic is a one-time event. The diagnostic device or diagnostic assay is used once at the beginning of the treatment. However, treatment in many cases is a multi-stage process with multiple procedures, which can be billed separately.

We can use wound care as an example. If a patient presents with a chronic leg ulcer, then in many cases, a doctor will use a Laser

---

[11]Popken B. (Aug 19, 2016) Martin Shkreli Weighs in on EpiPen Scandal, Calls Drug Makers 'Vultures,' NBC. Retrieved from https://www.nbcnews.com/business/consumer/martin-shkreli-weighs-epipen-scandal-calls-drug-makers-vultures-n634451

Doppler, toe pressure, or some other technique to diagnose the problem, which is a one-time event. Then, she will prescribe some dressing as a treatment. However, the dressing needs to be changed many times. It will involve multiple supplies and many hours booked by a doctor or nurse to change the dressing and monitor the progress.

Thus, the whole system has some skew toward treatment. Treatment is a financially more attractive option for healthcare. It also becomes a more attractive option for investors. In the absence of end product price elasticity, therapy becomes a much safer play for any investor.

9.5   Founder's Age

It is widely perceived that startups are for young people. Movies like "The Social Network" set expectations that all innovations are done by college-age youth.

However, the statistics of success tell us a different story. The probability of success increases with age.

It is even truer for life sciences startups. If you attend any life science startups' conference, you will be surprised with the average age of startups' CEOs.

In the recent study posted in the Harvard Business Review, the group of scientists from MIT[12] analyzed the US Census Bureau data with a focus on companies, which can be characterized as startups (e.g., were granted a patent, received VC investment, or operated in an industry that employs a high fraction of STEM workers). They found that the average age of high-tech founders falls in the early forties. They also considered "highest-growth" startups, which they defined as the top 1% of startups by growth. They found that patterns in all startups and high growth startups are similar and age groups 30–39, 40–49, and 50–59 has around 30%/30%/20% representation. 60-years old and older represent approximately 5%.

They also found that work experience plays a critical role, particularly if the experience is in the same narrow industry as their startup. Compared to founders with no relevant experience, those

---

[12]Azoulay P., Jones B. F., Kim J. D., and Miranda J., (2018) Research: The Average Age of a Successful Startup Founder Is 45, *Harvard Business Review*. Retrieved from https://hbr.org/2018/07/research-the-average-age-of-a-successful-startup-founder-is-45.

with at least three years of related work experience were 85% more likely to launch a highly successful startup.

As expected, the average age varied significantly between different industries. In software companies, the average age is 40, and younger founders are quite typical. However, other sectors such as oil and gas or biotechnology have higher average age, close to 47. They attributed the common perception about the founders' age to "disproportionate exposure to a handful of consumer-facing IT industries, such as social media, rather than equally consequential pursuits in heavy industry or business-to-business sectors."

They also found that the average age of the most successful (top 0.1% of startups) startup founders was 45 years.

Another finding of the study is that the probability of extreme success (top 0.1%) rises with age until the late fifties.

The authors of the study also pointed out that while such outliers as Bill Gates or Steve Jobs started their ventures in their twenties, the growth and most successful business decisions were achieved in their forties and fifties.

Thus, age is not considered a significant restriction or limitation. Older age is associated with a higher success rate. At a younger age, the lower success rate is compensated by more opportunities to start over.

Unfortunately, in addition to the public perception about founders' age, there are multiple indicators that VCs share the same preconception. For example, Paul Graham, a co-founder of Y Combinator, once mentioned that "the cutoff in investors' heads is 32... After 32, they start to be a little skeptical."[13]

While VCs may be outright wrong, and data support this notion, there is a plausible explanation for such bias. Any VC investment is a long-time endeavor. Founders and VCs are getting "married" for many years. Thus, it is conceivable to expect that VCs are looking for "teachable" entrepreneurs willing to follow their advice. Such as there is a perception that mental flexibility decreases with age and life experience, VCs may have a bias toward younger teams.

The MIT study authors proposed a more benign explanation for such a bias: "VCs are not simply looking to identify the firms with

---

[13]Rich, N. (May 2, 2013) Silicon Valley's Start-up Machine, *New York Times*. Retrieved from https://www.nytimes.com/2013/05/05/magazine/y-combinator-silicon-valleys-start-up-machine.html

the highest growth potential. Rather, they may seek investments that will yield the highest returns. It is possible that young founders are more financially constrained than more experienced ones, leading them to cede upside to investors at a lower price. In other words, younger entrepreneurs may be a better "deal" for investors than more experienced founders."

9.6   Market Size

The market size for the startup is of great importance. It is even more critical for MedTech startups, given their longer time to market.

As a rule of thumb, the following market size bins can be used (all numbers are annual sales):[14]

- If the market opportunity is less or around $1Mln, it is not a startup. It is a hobby instead. Bootstrapping is the best funding option.
- If the market opportunity is around $10Mln, then friends and family funding is the best funding option.
- If the market opportunity is around $100Mln, then angel investors may get interested.
- If the market opportunity is more than $1Bln, then you will get VCs' attention.

Note that we refer to the total market size. It can be very different from your beachhead market, which can be a small fraction of the total market.

So, the market size matters.

9.7   In-House vs. Outsourcing

Bringing the medical device to market requires efforts from various specialists. Besides traditional sales, marketing, operations, and finance, it also requires highly specialized professionals, ranging from hardware and software engineers to regulatory affairs to clinical coordinators to specialists on quality management systems. With so many activities and tasks, which need to be performed, the

---

[14]Walker, W. F., Creating Successful Academia Spin-Out, Virtual Medical Device Playbook, Feb 26, 2021.

natural question arises where to find all these people. It can be a particular challenge if your business is located not at a technology and MedTech hotspot.

Fortunately, multiple highly specialized consultants can help you complete any task or build a specific capacity or function.

Outsourcing (if it is done right) is a lower-cost option than building and maintaining an internal team. And in many cases, you don't need a full-time employee for a one-time task, e.g., implementing a QMS. A short contract to help with a specific task can be a much more optimal solution.

This discussion raises another important question: whether to build a specific capacity within the company or outsource this service?

The rule of thumb here is to develop in-house everything which is your core competency and mission-critical for the business and outsource everything else. Here I refer to outsourcing quite broadly. There are several other ways to get things done with minimal company involvement. For example, with the rise of SaaS, you can get an off-the-shelf product (e.g., for bookkeeping) that is priced per use that previously you had to get a license for. Or you can partner with somebody for sales. Thus, the rule of thumb can be extended to partnering, SaaS, or outsourcing nonessential items.

Core Technology

Your core technology is typically a key to your business success. Thus, in many cases, this function is retained within the business. It makes sense to build a team, which will develop and maintain your product. It is also prudent from an IP protection perspective. In this scenario, it is easier to maintain trade secrets.

Technology

However, not all hardware or software development is necessarily your core technology. For example, if your product is a standalone device, which can work with a smartphone, then the smartphone app development can be an auxiliary function. In this case, it can be easily outsourced at the early stages of the company. However, you may reconsider this decision at a later stage when you have a bigger team.

Clinical Affairs

Clinical affairs, in many cases, are the core function of the MedTech startup. It would help if you started building relationships with the medical world at very early stages. Thus, having somebody with a medical background within the team can be critical to success. One option here can be having a Clinical Advisor.

However, in many cases, it is not necessary to go beyond one clinical expert and build extensive clinical expertise in-house. With the proliferation of Contract Research Organizations (CROs), many functions during clinical phase can be outsourced, including recruiting test sites and participants.

Regulatory Affairs

Some functions are difficult to build in-house. Regulatory affairs can be such an example. The regulatory environment is a constantly evolving field, with multiple regulations introduced every single day. Especially difficult is to keep pace with that at several jurisdictions simultaneously. And not many specialists have expertise in several markets. And it is always helpful to have somebody familiar with premarket submission while preparing 510(k) or PMA. Thus, hiring a regulatory consultant can be a very prudent approach.

Reimbursement

The same logic applies to reimbursement. Reimbursements are a highly specialized area, which requires significant expertise and connections within the industry. It can be challenging to build this function in-house.

Thus, regulatory affairs and reimbursements are natural choices for outsourcing, particularly at the venture's early stages.

QMS Implementation

Another ideal candidate for outsourcing is quality management system implementation. It is a one-time, highly specialized task. I would recommend hiring an industry expert to implement QMS, develop required forms and procedures, and train the internal team. Then, the internal team will be able to maintain QMS going forward.

In addition to MedTech-specific functions, many other business functions can also be outsourced.

Finance

You can easily find contractors for bookkeeping and accounting services.

At the early stages of the venture, there is no need for a full-time CFO. Instead, you can hire a fractional CFO if you need such services. However, getting upskilled in financial literacy is essential for all founders. You do not need to be an accountant, but understanding how to read financial statements and understand key concepts will help you make better decisions as a founder. Liquidity is one of the top reasons for existing businesses to go bankrupt; this is even more critical for startups.

Marketing

As MedTech is a regulated industry thus, the premarket activities are quite limited. Moreover, the long development, clinical, and regulatory approval phases make marketing even more challenging. Therefore, in many cases, there is no need for a full-time marketing specialist. In many cases, this function can be outsourced to contractors for a fraction of the cost.

Sales

As you may notice from the previous discussion, sales cannot be considered as a primary function in MedTech startups at the early stages. The product is far from the market, and multiple restrictions limit premarket communications. Thus, the question about sales can be deferred to a later time. However, it should be mentioned that using channel sales and distributors in the MedTech world can be a much simpler solution than building a large in-house sales team. So, although hiring a salesperson may be premature, building and understanding the sales strategy is rarely a waste of time.

Manufacturing

If you are not a manufacturing expert, then building the manufacturing function in-house can be a challenge. In most cases, the facility (particularly where the final assembly happens) needs to be ISO 13485 certified. Thus, consider outsourcing to an ISO 13485-certified contract manufacturer.

9.8   Location... Location... Location

Startups share one important trait with real estate: location is very important. Recently, I have heard, "You can't do in Kansas." With all due respect to the great state of Kansas, it may hold true in the startup world. It is hard to build a viable MedTech startup in general. It can be even harder to do it outside of traditional MedTech clusters.

There are several reasons for it. Firstly, it is connected to resource availability. Large MedTech and BioTech centers like New York, Boston, San Diego, North Carolina, Medical Alley (Minnesota), or Toronto have multiple resources available. It includes research infrastructure, industry support, and a skilled workforce.

As we discussed through the book, any MedTech startup requires talents from many disciplines. In large MedTech clusters, it can be relatively easy to find the talent needed. Moreover, the mindset of many professionals is startup-oriented. Many of them are ready to take a calculated risk and jump into the startup world. Furthermore, these locations have multiple universities with relevant programs. However, it can be hard to find a required specialist outside of these MedTech clusters.

Large centers also have required research and clinical infrastructure. For example, they have large research hospitals, which are comfortable with trying new technologies. Thus, it can be much easier to find a Clinical Advisor or perform verification in a clinical environment.

Finally, it is connected to investor availability. As we mentioned before, many investors prefer startups in their backyards. Investment in a MedTech startup has a long-time horizon, so logistically, it is easier to have it nearby. It allows great saving of time on travel (for example, attending board meetings). Typically, investors are OK with many of these large MedTech and BioTech clusters; however, they may have some reservations about a long-term commitment to off the beaten path.

Obviously, residing in large centers has its disadvantages. They are typically associated with much higher salaries. In addition to that, typically, the office/facility rent is also higher, which increases overall overhead. However, in many cases, these costs are still reasonable trade-offs for the ability to have untapped access to resources.

However, these things may change in the future. Several trends, which will be discussed in the next chapter, can level and democratize the field.

An even bigger question is in which country the startup is located. While things are appeared democratized and globalized cross-border, in practice, it is much more complicated. For example, let's consider investment availability. It is well known that the United States has the largest VC pool in the world. Moreover, these are experienced investors who understand risks very well. Thus, getting funding in the United States is simpler than in many other places. American investors are obviously fine (with some caveats we discussed several paragraphs above) with the US startups (preferably incorporated in Delaware). Some United States investors are OK with Canadian startups. An even smaller fraction of them will be fine with EU startups. However, going further, things may get really complicated.[15] It is one thing when the US startup has a Ukraine-based CTO or Russia-based technology team. However, if the same startup is incorporated in Russia or Belarus, its fundability from the US investor perspective will be completely different. Investors prefer well-known and well-functioning legal systems.

---

[15]It refers mostly to (a) traditional VCs and (b) MedTech. VC arms of multinationals with footprint in many countries (Corporate VCs) may be comfortable with investing in other startup hotbeds like Israel and India.

Chapter 10

Silver Lining

In the previous chapters, I tried to explain to you the intricacies of MedTech startups. My goal was to equip you with tools and prepare for MedTech entrepreneurship. So, I hope you will be prepared and able to analyze the odds if the opportunity arises. However, it is not all doom and gloom. There is a silver lining. In this very short chapter, I will demonstrate that scientists can significantly improve this game's odds.

We possess several skills and traits, which are far less common than we think and can be particularly helpful in MedTech entrepreneurship.

Analytical skills. All scientists have analytical skills honed by years of research and trials and errors. We can observe, make conclusions, and change our behavior accordingly. We can decompose complex problems into simpler ones and solve them sequentially.

Scientific method. Scientists use a scientific method routinely. Formulating a hypothesis, planning and conducting experiments to verify this hypothesis, and modifying the theory based on experimental results are part of our daily routine. However, as I mentioned before, the scientific method is not limited to academia anymore and started getting traction in business recently. Nowadays, the scientific method became the pillar of the Lean Startup

Bringing a Medical Device to the Market: A Scientist's Perspective
Gennadi Saiko
Copyright © 2022 Jenny Stanford Publishing Pte. Ltd.
ISBN 978-981-4968-25-6 (Hardcover), 978-1-003-31221-5 (eBook)
www.jennystanford.com

methodology. Such as it is in our genes already, we are well equipped to apply it in all aspects of entrepreneurship.

Teachability and ability to learn. Entrepreneurship requires a lot of continuous learning. Fortunately, scientists are eager to learn new things, which is a part of our daily practice.

Perseverance. Entrepreneurship, and particularly MedTech entrepreneurship, requires a lot of perseverance. As luck would have it, it is another scientist's common trait. We are well familiar with working late nights, running many unsuccessful trials, harsh critique from peers, multiple rejections from journals and granting agencies. All this experience builds our stamina, which will be particularly helpful in the startup environment.

Credibility. Scientists have credibility by default. Even though scientists turned into entrepreneurs may meet some suspicion from their colleagues, they are still scientists to the general public, investors, and other counterparties. It can be crucial in gaining credibility from investors, partners, and customers.

Scientific infrastructure. Often we have space (lab and office) and equipment to run experiments. Even if we don't have it, we know where to find it. Oftentimes, this equipment can be utilized at significantly discounted prices. So, access to university infrastructure can be a critical factor of success.

In addition to that, academia is eligible for multiple sources of grant funding. Given the long process of bringing the MedTech project to the market, academia is the best place to nurture it.

We also have access to top subject matter experts in the field.

Access to talents. We have access to large talent pools. Often, we work with very talented students at the beginning of their spectacular career journeys. Moreover, we understand their motivation. In many cases, their commitment and length of stay (e.g., graduate students) are highly predictable.

This list may go on and on. However, I guess you grasped the idea. We, as scientists, are uniquely equipped to succeed in the "MedTech" world. Note that the keyword in this statement is MedTech. Our experience and access to resources and talents are very relevant to the MedTech world. However, they may be less relevant or perceived to be less relevant to consumer goods, FinTech, and other types of startups.

Fortunately, there is no need to look at other sectors. MedTech is a great place to be. It is a very rewarding experience. And with changes to our society, there are good chances that the healthcare system is ripped for changes. Here is a list of several reasons why to choose MedTech entrepreneurship.

Solving real problems. While developing another dog walker app will most likely bring you much more money than a MedDev project, MedTech is vital to our society.

Think of any car accident victim or our society in general as a result of COVID-19 pandemic. We rapidly slide from high levels in Maslow's hierarchy (self-actualizations and esteem) almost to the bottom, to safety needs. Thus, our work is critical to society in general.

Startup gurus often refer to customer needs as "customer pain." In our case, it is not a figure of speech. We deal with real patients who have real pain.

Saving lives and improving patients' quality of life. It is a potent motivator.

Let's consider wound care as an example. It is known that amputees have a lower life expectancy. For example, the 5-year mortality rates for patients with diabetic foot ulcers and after amputation are 30.5% and 56.6%, respectively.[1] With the current pandemic of diabetes, when almost 10% of the population will develop diabetes during their lifetime, it results in millions of amputations annually and significantly reduces life expectancy for a considerable number of people. Just imagine if you can improve outcomes and life expectancy after amputation just slightly. How many person-years of human lives will be saved? Or even better, if you could prevent amputations just slightly, by several percent. In this case, hundreds of thousands of people will have a much better quality of life and a much higher life expectancy.

---

[1]D. G. Armstrong, M. A. Swerdlow, A. A. Armstrong, M. S. Conte, W. V. Padula, S. A. Bus (2020). Five-year mortality and direct costs of care for people with diabetic foot complications are comparable to cancer. *J Foot Ankle Res*, 13(1): 16.

Working with amazing people. We have the privilege of working with healthcare professionals. They are amazing people. They are smart and passionate, the true unsung heroes of modern society.

Changing healthcare to better. Healthcare is a relatively conservative industry. While there are many fantastic healthcare innovations; however, at its core, it is a very conservative field, which tries to maintain its status quo.

For example, in primary healthcare, many exams are performed by the naked eye. Obviously, it is a highly subjective method, which depends on the doctors' experience. So, to some extent, it is more art than science.

We, as MedTech entrepreneurs, can change it. We can bring quantitative objective methods. There is a lot of resistance from the industry. Moreover, there are significant regulatory hurdles, particularly for AI. But, with more and more innovations coming, the field is gradually changing already.

Hopefully, by developing and bringing new technologies to the market, we will be able to level the field, objectivize diagnostics, and democratize access to world-class healthcare to every citizen.

So, there is a lot of work ahead.

Final Thoughts

Finally, we are at the end of the book. I hope I have not scared you. I want to finish this book on a positive note. MedTech entrepreneurship is a gratifying experience. Saving lives and working with amazing people are potent motivators. So, it is definitely worth trying. I wish you good luck in your journey.

The End.

Instead of Afterword

The whole book before this point was written under the assumption of the pre-COVID status quo. However, COVID-19 pandemic and several other recent trends may change this status quo significantly.

Obviously, I do not have a crystal ball to foresee the future. In general, such a prediction can be wildly inaccurate. Moreover, recent events (like COVID-19 pandemic) show that real life is much more vivid and creative than human imagination (who may predict devastating events of 2020 even at the beginning of 2020?). However, it is still worth trying to outline several important trends other than some obvious ones as rise of AI-based technologies.

As of the time of writing (the end of 2020 – beginning of 2021), the world is still in the middle of the COVID pandemic. This pandemic will certainly induce significant changes to many aspects of human life, including healthcare. However, right now, most of these changes are unclear. Currently, we wear masks, countries are at various stages of shutdowns and lockdowns, international travel is significantly restricted, and people are urged to obey social distancing. Hopefully, we will achieve the required immunity levels with vaccinations, and life will start returning to normal. However, what will this normal be? While there are not many visible changes that can be predicted right now, it is quite apparent that some tectonic shifts are in the making.

We will start our discussion with several changes, which the COVID pandemic can cause. After that, we will discuss several general trends, which are not directly related to COVID.

Indeed, the proposed list is far from being complete. Most of the trends discussed will likely have mid and long-term effects and not short-term effects. For example, it is hard to expect that most of these factors will significantly impact the next three years. However, as we discussed, a MedTech venture is a long-time project. It is better to think strategically and keep these factors in mind.

Virtual Meetings

If we need to pick one of the most obvious consequences of the COVID pandemic on business, it is a proliferation of virtual meetings. Many aspects of business and education moved online. And it is not clear which aspects will revert back to in-person meetings.

Obviously, most of the education (particularly junior, middle, and high school) will revert at the earliest possibility. Children need to communicate in person to develop social skills. However, the pathway for business is less obvious.

The whole pre-COVID business world relied heavily on meetings in person. All business development relied on traveling and meeting in person. Sales representatives and sales executives spent most of their time hopping across the country. For example, all morning and evening flights between major cities in the United States and Canada were packed with business travelers.

While it was particularly true for medium and large businesses, it was true for startups as well. If you fly from any major North American city to San Francisco, you may notice that a significant part of travelers is startups' CEOs who fly to Silicon Valley to meet with investors.

Most likely, it is going to change dramatically. Right now, all these activities moved online. After the initial shock in March–April 2020, most business activities in the startup world started recovering. Now, most meetings happen online, and all business sides are getting more comfortable with it. The new virtual reality presents significant savings in terms of time and money. Thus, it will be partially adopted in the after-COVID world, even in the short run.

Such changes will have immense impacts on many aspects of our life.

Probably, the travel industry will have the most significant impact. The whole industry took a substantial hit in 2020; however, this trend will most likely persist. Business travel is the backbone for many segments of the travel industry. For example, the business class was the primary revenue driver in airlines. Hotels are also very dependent on business travelers, who easily spend the company's money on lodging, food, and drinks. If business travel revenue dries up, it will reshape the whole industry, most likely resulting in multiple bankruptcies and consolidations. Most likely, it will result in fewer options and higher prices.

However, for MedTech entrepreneurs, it provides significant advantages.

Firstly, as we mentioned before, it provides significant time and cost savings, which is particularly important for cash-strapped entrepreneurs. However, benefits may go well beyond that.

Zoom meetings democratize a lot of things. They are much easier to book than in-person meetings. For example, let us imagine a meeting between decision-makers of a healthcare system and an aspiring startup. As you may imagine, the healthcare decision-makers are quite busy, so these meetings are book for months in advance. Moreover, such a meeting is quite a commitment, especially if the visitors fly from another part of the country. With Zoom meetings, everything becomes less complicated. An executive can attend this meeting even between rounds of golf. So, hopefully, virtual meetings will significantly simplify business development.

Finally, virtual meetings may level the investment field. As we mentioned before, many investors prefer to invest in their "own backyard." In this case, you do not need to fly for multiple hours for a board meeting. However, if in the future board meetings will take place online, this factor becomes less critical. Thus, perhaps investors will broaden the geographical boundaries of their portfolio companies.

Virtual Workforce

COVID pandemic changed significantly human perception of the work environment. While many companies experimented with working from home arrangements for many years, the COVID pandemic may forever change the work environment.

Working from home and flexible working hours were perks, which innovative companies offered to their employees to attract and retain the brightest. However, in most cases working from the office was still predominant mainstream.

However, this paradigm is likely to be shuttered. With some industries working completely remotely since March 2020, working from home may become mainstream for all creative workforce.

The primary driver here will be employees' perception. While many of them (especially with young kids) had a hard time adjusting to the new reality, many people understood its advantages in the course of the year. So, it will be virtually impossible to force these people to work in the office from now on. Most likely, a hybrid environment will become the mainstream. It can be the employee's discretion to work from the office or home, but the office environment can no longer be mandated. Most people will not tolerate it and leave the company for more flexible arrangements. I would compare it with corralling cats. Creative workers are entirely independent, so it will be virtually impossible to achieve.

It will impact the business in several profound ways.

Firstly, companies may be able to reduce overhead significantly. There is no need to lock into expensive long-term leases on prime real estate. It will have a devastating impact on commercial real estate and downtown businesses, but it is very beneficial for cash-strapped startups.

Secondly, it may level the field. The location of the startup will be less critical. In the pre-COVID era, startups in large tech clusters like Silicon Valley, NYC, Toronto, or North Carolina had significant advantages over smaller cities' peers. For example, they had access to large workforce pools. At the same time, startups in non-primary locations had a hard time attracting talents. Now, this field will be leveled.

Thirdly, it will lead to a redistribution of the population. With no necessity to live in tiny apartments in an expensive downtown, many people moved already or will move to smaller towns and rural areas, which will create interesting healthcare patterns, which we will discuss in the following sections.

Telehealth

Even in 2019, telehealth was perceived as a remote opportunity. While technology had been here for a long time, the medical community definitely did not favor telehealth, which threatened to reduce fees for service. Thus, even assuming that some policymakers favored broader access to telehealth, it was virtually no starter due to unified pushback from the medical community. Consequently, Telemedicine was reserved (at least here, in North America) for quite limited applications like remote and indigenous communities.

This status quo was changed virtually overnight with the arrival of COVID. The scope of telehealth services coverage (reimbursement codes) was extended through the whole primary care. Many patients experimented with these services and were quite satisfied. While some payers claim that this coverage will be rolled back in the post-COVID world, it is not likely to happen. Now, they will have a massive pushback from patients if they attempt to do so. Patients got accustomed to these services. They do not necessarily want to travel a long way to get ten minutes of doctor's attention. Not all visits can be moved online, but some percentage will likely stay there.

This market sentiment is confirmed by venture funding. In 2021 Q1 MobiHealthNews reported[2] on whopping 99 digital health deals equaling $7.1 Bln in funding. This number exceeds significantly the $2.9 billion reported in 2020's Q1.[3] A direct-to-consumer virtual care company Ro secured $500 million in funding.

The other factor contributing to telehealth adoption will be the redistribution of the population, which we mentioned in the previous section. With young families moving into small towns and rural areas, they will create a push for medical services. While family physicians may gradually move to fill the gap, it is not likely to happen with specialists. But even with family physicians, it will be a significant lag. Thus, telehealth and remote patient monitoring may help to fill the gap at least temporarily.

Finally, the healthcare authorities probably noticed how unprepared the health system was for such an event as the COVID outbreak. With the limited built-in extra capacity, the healthcare systems are not flexible enough to cope with such challenges. Telehealth and remote patient monitoring allows rapid and scalable response even when the traditional healthcare system is overwhelmed and shut down. Thus, hopefully, they will be nurtured and preserved as backup solutions.

Failure of current healthcare models

Probably the biggest lesson from the COVID pandemic is that the public health systems in many countries are not prepared for such challenges.

---

[2]MobiHealthNews, Investors pour $7.1B into digital health investments during Q1 2021. Retrieved from https://www.mobihealthnews.com/news/investors-pour-71b-digital-health-investments-during-q1-2021

[3]Note that Q1 2020's figures were not affected yet (at least significantly) by the COVID pandemic.

Countries where the government took an active role in handling the pandemic (China, Korea, Japan, Taiwan, Australia, and New Zealand) successfully contained the epidemic.

However, countries, where the government took a more passive role (particularly the US, Mexico, and Sweden) encountered significantly higher per capita mortality than nearby and peer countries.

Canada represents a very interesting example of the government response. While, in general, Canada had several times lower per capita mortality than the nearby US, the mortality and new cases distribution varied significantly between different provinces. The best kept Canadian secret is an "Atlantic bubble." Four Canadian provinces (Nova Scotia, PEI, New Brunswick, and Newfoundland and Labrador) implemented strict rules for traveling outside the shared borders (a two-week mandatory quarantine). They managed to have only isolated cases of COVID and a small number of deaths. However, provinces that implemented fewer restrictions (like Quebec, Ontario, and Alberta) had almost overwhelmed their healthcare systems.

The lesson here is that a profit-oriented healthcare system is not ready to handle such events. To increase profits, private healthcare providers reduced overhead by building only limited extra capacity, capable of handling only seasonal fluctuations. This system is not capable of handling major events, like the COVID-19 pandemic.

It is hard to expect that private businesses learn this lesson and will start building that extra capacity. Somebody else needs to think strategically and prepare for such kind of events. The government is the only player who is capable of doing it.

Thus, it is reasonable to expect that there will be some significant changes to healthcare systems in countries like Canada and the United States. However, these are political issues. Thus it is hard to expect that it happens immediately, but it almost certainly may occur in the midterm (3–5 years).

For example, in Canada, according to the National Institute on Aging (NIA) at Ryerson University, by January 5, 2021, long-term care (LTC) and retirement homes reported 11% of the Canadian totals of COVID-19 cases and 73% of total deaths[4] while they account for less

---

[4]Loprespub (Jan 5, 2021), Long-Term Care Homes in Canada—The Impact of COVID-19, *Hillnotes*, Retrieved from https://hillnotes.ca/2020/10/30/long-term-care-homes-in-canada-the-impact-of-covid-19/

than 1 percent of the overall population, and 7 percent of all seniors. Thus, it is not hard to see that the current model is unsustainable and up for a complete overhaul.

In addition to the health system overhaul, some changes need to be made to preparedness. Currently, if a company works on a compound against chemical or biological threats, it will most certainly be out of favor among investors. It does not represent an exciting investment opportunity with no mass-market potential and the only customer as the US Army or DARPA. For example, according to a recent report,[5] Big Pharma does not have any projects in its pipelines to address 10 out of 16 infectious diseases identified by the World Health Organization as the most significant public health risk. However, the year 2020 shows that it is crucial and needs to be done. The government is the only player, which can finance such public goods. So, hopefully, such preparedness agencies as BARDA in the United States will increase funding for such companies.

K-Shape Recovery

The total effect of the COVID-19 pandemic on the global economy and individual countries is unknown. In 2020, most OECD countries (with China as a notable exception) posted a drop in GDP, caused predominantly by a significant plunge during the initial shutdown in Q2. Since then, economic indicators started recovering, and part of the initial job loss was recouped.

However, the story is far from over. Currently, all large countries (with Russia as a notable exception) pump a significant amount of money to companies and people.

In April 2020, it was assessed that the United States had committed more than $6 trillion to arrest the economic downturn from the pandemic.[6] As of October 1, 2020, $2.6Trln was spent on stimulus spending and $900Bln in Stimulus Tax Relief. These numbers do not include the December 27, 2020 stimulus (The Consolidated Appropriations Act), which included $868 billion of

---

[5]J. Kollewe (Jan 26, 2021), Pharmaceutical giants not ready for next pandemic, report warns, *The Guardian*, https://amp.theguardian.com/science/2021/jan/26/pharmaceutical-giants-not-ready-for-next-pandemic-report-warns

[6]A. van Dam (Apr 15, 2020), The U.S. has thrown more than $6 trillion at the coronavirus crisis. That number could grow, *The Washington Post*, https://www.washingtonpost.com/business/2020/04/15/coronavirus-economy-6-trillion/

federal support. On March 11, 2021, the huge $1.9 Trln coronavirus relief package was signed into law.

As of October 2020, most G20 countries spent more than 7% of their GDP on COVID-19 fiscal stimulus (Japan: 21.1%, Canada: 16.4%, United States: 13.2%, Germany: 8.9%, China: 7%, India: 6.9%).[7]

Thus, currently, all major economies are under steroid injections. However, it is not clear what happens when this help will be withdrawn.

The preliminary data indicate that stimulus money was misallocated. In most cases, consumers were scared and have not increased their spending. Thus, stimulus money did not go into manufacturing (again with China as a notable exception, which spent significant money on infrastructure) but rather into savings. Thus, instead of real economy stimulus money was pumped into the stock market and real estate and propelled it to new heights.

At the same time, the small businesses are in perils. If you visit any shopping street in Toronto or its affluent suburbs, it is easy to notice how many stores and restraints have been closed permanently. According to some estimates, up to 1/3 of small businesses will not survive COVID-19 pandemic.

Economists refer to it as a K-shape recovery, where various population strata have different outcomes: wealth growth for the wealthy and impoverishment for everybody else.

The net result of such recovery is entirely unclear. For example, it can trigger further recession or political unrest.

The impact on healthcare and healthcare startups is also hard to predict, but it will probably be negative. In the short term, the availability of investments may increase. However, due to other priorities (like fighting a recession), healthcare's required changes can be postponed. Investment also will take a hit in the case of recession.

Regulatory Environment in the European Union

In this section, we will talk about by far the smallest global factor on our list. However, it still may have implications for MedTech entrepreneurs, particularly in the short and midterms.

[7]Statista, Value of COVID-19 fiscal stimulus packages in G20 countries as of March 2021, as a share of GDP. Retrieved from https://www.statista.com/statistics/1107572/covid-19-value-g20-stimulus-packages-share-gdp/

I am referring to the new European Union Medical Device Regulation (MDR), which replaces MDD from May 2021. The MDR implementation is a long-awaited event, which will have significant implications on the MedTech world. The deadline for implementation was pushed already once in 2020, and it will not be surprising if it will be pushed one more time due to COVID-19. However, sooner or later, its implementation is imminent, and changes and consequences will be drastic.

The reason for that is the following. The previous European MedDev regulations (Medical Device Directive or MDD) were relatively liberal. For example, the scope was to demonstrate the device's safety, which is a significantly lower barrier than demonstrating safety and effectiveness, as required by the FDA.

Thus, getting regulatory approval in the European Union was a significantly more straightforward, faster, and cheaper task than getting regulatory approval in the United States (see Chapter 5, Timelines and Capital for details). Notably, it was true for novel and high-risk devices. And given the size of the market (446Mln people, which is almost 40% larger than the United States), it is obvious why relatively large portions of entrepreneurs choose to go to the European Union first to lower their costs and time to market. As a result, the European Union has a broader spectrum of novel, experimental diagnostic and therapeutic devices.

The MDR will change it completely. The most apparent change is a requirement for clinical data, which was not required in the MDD world. In addition to that, from many aspects, this regulation is stricter than in the United States.

For example, we can consider the so-called Rule 11, related to software as a medical device (SaMD).

"Under Rule 11 of the MDR, pretty much any SaMD that provides clinical information—such as information used for making decisions for diagnosis or treatment" will be classified as a Class IIa medical device or higher.

Under the existing Medical Device Directive (MDD) rules, the majority of this type of SaMD would have been Class I, the lowest risk category."[8]

---

[8]P. Maquire (2020). MDR Rule 11: What the change means for medical device companies, Retrieved from https://www.s3connectedhealth.com/blog/mdr-rule-11-what-the-change-means-for-medical-device-companies

This 'up-classification' means MedTech companies with digital health solutions will have to comply with a range of new rules and requirements.

The MDR implementation will have several important implications for MedTech entrepreneurs.

Firstly, the new procedures have not been tested yet on a large scale. Their implementation can cause significant delays for regulatory approvals in the short term.

Secondly, most MedTech entrepreneurs will most likely lose the possibility of "easy entrance" to markets through the European market.

Thirdly, existing companies, which have their products on the European Union market, will need additional regulatory work. The good news is that this work can be postponed by three years according to the transitional arrangements provided under MDR Article 120(3). This provision allows devices certified under MDD before the May 26, 2021 deadline can remain on the market until May 26, 2024, by which time they will need to be resubmitted under MDR. The bad news, it will cause an avalanche of re-certifications closer to 2024 and significant delays.

Finally, all consequences of the Brexit on MedDev regulations are not very clear. It is known that the UK MHRA has decided that it will implement its own regulatory scheme for medical devices and will not follow the EU MDR. The UK will be issuing its own UKCA (UK Conformity Assessed) mark to replace CE Marking. CE Marking will not be recognized in the United Kingdom after June 2023. Thus, as the minimum, entrepreneurs will have one more set of market regulations to comply with. This will add costs and complexity. On the global scheme of things, most likely, instead of the current five (the United States, European Union, Canada, Australia, and Japan), we will have six (five + the UK) market regulations, which regulators from other countries will benchmark.

Pace of Globalization

We will finish the book with probably the most fundamental recent trend: the slowing pace of globalization.

Globalization is probably the most important global phenomenon of the end of the 20th century and the beginning of the 21st century. It interconnected multiple economies and lifted from poverty many countries, particularly in East and South-East Asia.

However, some indicators show that this phenomenon halted or at least its pace will be changed in the near future.

The most apparent indicator of these changes is the rift between the United States in China, which began in 2017. The Trump administration started the trade war, and China reciprocated it. As a result, multiple aspects of global supply chains started readjusting.

It is not clear whether these tensions will be smoothened in the post-Trump era. However, it is quite clear that the business took a note, and it will start readjusting its strategies.

The cracks in globalization trends were particularly on display during the COVID pandemic. While it was expected that it is a global problem and approach should be global, many countries took a very individualistic stance.

It was started on April 2, 2020, when Trump's administration banned the export of medical supply to Canada and high jacked medical supply shipments to Germany and France. At the same time, Germany rejected helping Italy,[9] while other countries like Russia, China, and Ukraine stepped in.

The history repeated itself in December 2020–January 2021, when first vaccines became available and rich countries rushed to secure stocks for themselves, depriving less fortunate COVID-battered countries like Ukraine or South Africa of life-saving supplies. Even Canada was pushed on the backburner and received an unproportionally small fraction of vaccines. More recently, on January 26, 2021, the European Union mulled export control of vaccines produced there until its market is saturated[10] (Pfizer's vaccine for Canada is made in Belgium).

Most likely, these nationalistic decisions will not result in some sudden changes. Moreover, all countries will be vaccinated a little bit sooner or several months later. Still, it is plausible that politicians in many countries took a note, which may result in re-aligning

---

[9]K. Adler (Apr 2, 2020). Coronavirus outbreak eats into EU unity, *BBC News.* Retrieved from https://www.bbc.com/news/world-europe-52135816

[10]M. Peel, S. Fleming (Jan 27, 2021) Vaccine export rules: what is the EU proposing? *Financial Times.* Retrieved from https://www.ft.com/content/875f6a1f-e740-4a4d-b203-24e43c09b654

priorities and alliances in the medium and long term.

How may it impact healthcare and MedTech entrepreneurs? The impact can be quite profound. Firstly, it may impact supply chains and make things like prototyping and product development much slower and costly. Secondly, it may affect the overall trend on trade and regulation harmonization. For example, countries may try to protect their own manufacturers and inhibit access of foreign players.

Appendix A

Premarket Notification [510(k)] Process

Such as most medical devices in the United States fall in the category of the Class II medical devices, the Premarket Notification, or 510(k), is of primary interest for many applicants. That is why we will devote this section to some specific aspects of the 510(k) process. We will start with a general overview of the process, including its timelines. Then, we discuss various types of 510(k) submissions. Finally, we will discuss the substantial equivalence and predicate devices, which are the core elements of the 510(k) process.

A.1 510(k) Workflow and Timelines

Under 510(k), a manufacturer must demonstrate to the FDA that their device is substantially equivalent (as safe and effective) to a device already on the market, which is not subject to Premarket Approval (PMA).

While Premarket Notification sounds like a notification, it is typically a multi-stage and multi-month review process (see Fig. A.1). Two primary phases of the 510(k) review are the Acceptance Review and Substantive Review. To get the device cleared, an applicant collects data about their device's safety and efficiency and submits the 510(k) application form to the FDA. A user fee payment should accompany the submission. If the user fee payment and the valid filing have been received, the FDA issues an Acknowledgement Letter with the date and 510(k) number and proceeds to the Acceptance Review. The Acceptance Review has to be conducted within 15 days and may result in one out of three outcomes:

- the 510(k) was accepted for Substantive Review; or

- the 510(k) was not accepted for review; or
- the 510(k) is under Substantive Review because FDA did not complete the Acceptance Review within 15 calendar days.

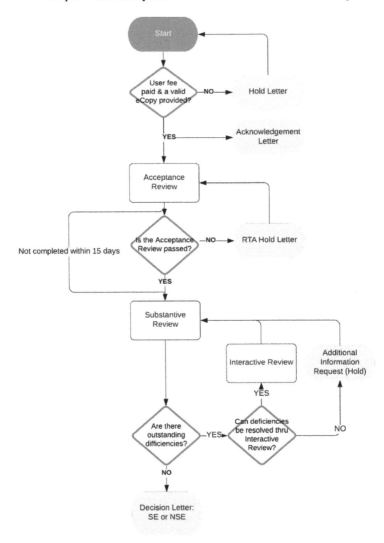

Figure A.1 Major steps and outcomes of the 510(k) review process.

If the submission was considered refused to accept (RTA), it is placed on hold. The applicant has 180 calendar days to address the deficiencies cited in the RTA Hold fully. If this is not done, the 510(k) is considered withdrawn and deleted from the review system. If the 510(k) is deleted, the 510(k) applicant will need to submit a new, complete 510(k) to pursue FDA marketing clearance for that device.

Once passed the Acceptance Review, a 510(k) proceeds to the Substantive Review. During Substantive Review, FDA conducts a comprehensive review of the 510(k) submission and communicates with the applicant through so-called Substantive Interactions. The FDA targets the Substantive Interaction within 60 calendar days of receipt of the 510(k) submission.

The typical Substantive Interaction outcome is binary. The FDA either proceeds with resolving any outstanding deficiencies via Interactive Review; or places the submission on hold with an Additional Information (AI) request.

If deficiencies are minor, FDA may choose to go through an Interactive Review process. During Interactive Review (the FDA targets 90 calendar days), the FDA may request additional information from the applicant. Typically, during Interactive Review FDA communicates with the applicant through phone and emails.

If deficiencies are more substantial, FDA may choose to send an AI Request, which places the submission on hold. The applicant has 180 calendar days from the AI Request date to submit a complete response to the AI Request. If the FDA does not receive a complete response to all deficiencies in the AI Request within 180 days of the AI Request date, the submission is considered withdrawn and deleted from the FDA review system.

The Substantive Review results in the 510(k) Decision Letter. The Decision Letter's primary outcome is determining whether the submission is substantially equivalent (SE) or not substantially equivalent (NSE).

A 510(k) that receives an SE decision is considered "cleared." FDA adds the cleared 510(k) to the 510(k) database, which is updated weekly.

FDA targets to issue the Decision Letter within 90 calendar days since the Acknowledgement Letter (excluding any days where the submission was on hold). These timelines are imposed by the Medical Device User Fee Amendment of 2012 (MDUFA III) as performance goals for the 510(k) process. However, suppose FDA does not reach a MDUFA decision within 100 FDA days (i.e., 10 days after the MDUFA goal). In that case, the FDA will issue a Missed

MDUFA Communication, a written feedback to the applicant, which includes the major outstanding review topic areas or other reasons preventing the FDA from reaching a final decision, with an estimated date of completion. These topics will be discussed later in a meeting or teleconference.

A.2 Types of the 510(k) Submission

There are three types of Premarket Notification 510(k) submissions: Traditional, Special, and Abbreviated.

A.2.1 Traditional 510(k)

Most new devices initially go through the Traditional 510(k) pathway. Any significant change to the device (significant change or modification in design, components, method of manufacture, or intended use) will require submitting a new 510(k). However, some device changes may be implemented without the submission of a new 510(k).

The Traditional 510(k) target timeframes are 90 days.

The Special 510(k) and Abbreviated 510(k) submission types can be used when a 510(k) submission meets specific criteria.

A.2.2 Special 510(k)

Device manufacturers may choose to submit a Special 510(k) **for changes to their own existing device** if "the method(s) to evaluate the change(s) are well-established, and when the results can be sufficiently reviewed in a summary or risk analysis format." For example, a Special 510(k) most likely will not be appropriate for devices that manufacture a biological product at the point of care because there would likely be no well-established method to evaluate such changes and/or the performance data would not be reviewable in a summary or risk analysis format.

On the other side, design or labeling change(s) to an existing device (including certain modifications to the indications for use) may be appropriate for a Special 510(k) when:

- "The proposed change is submitted by the manufacturer legally authorized to market the existing device;

- Performance data are unnecessary, or if performance data are necessary, well-established methods are available to evaluate the change; and
- All performance data necessary to support substantial equivalence can be reviewed in a summary or risk analysis format."[1]

The FDA generally reviews Special 510(k) submissions within 30 days of receipt

A.2.3 Abbreviated 510(k)

In certain circumstances, the applicant may submit an Abbreviated 510(k). The Abbreviated 510(k) filing relies on one or more:

- FDA guidance document(s), or
- Demonstration of compliance with special control(s) for the device type, or
- Voluntary consensus standard(s).

An Abbreviated 510(k) submission must include the required elements of the Traditional 510(k). However, in an Abbreviated 510(k) submission, the applicant provides a summary report on the use of guidance documents and/or special controls or declarations of conformity to the FDA's recognized standards. An Abbreviated 510(k), which is based on the general use of a voluntary consensus standard, should include the basis of such use and the underlying information or data that supports how the standard was used.

The FDA targets 90 days for the review of the Abbreviated 510(k).

Traditional and Abbreviated 510(k) submissions are fairly extensive documents. FDA recommends including the following sections:[2]

"(1) Medical Device User Fee Cover Sheet (Form FDA 3601)

(2) Center for Devices and Radiological Health (CDRH) Premarket Review Submission Cover Sheet (Form FDA 3514)

(3) 510(k) Cover Letter

(4) Indications for Use Statement (Form FDA 3881)

---

[1]FDA (2019). The special 510k program. Retrieved from https://www.fda.gov/regulatory-information/search-fda-guidance-documents/special-510k-program

[2]FDA (2019). Format for Traditional and Abbreviated 510(k)s. Retrieved from https://www.fda.gov/media/130647/download

 (5) 510(k) Summary or 510(k) Statement

 (6) Truthful and Accuracy Statement

 (7) Class III Summary and Certification

 (8) Financial Certification or Disclosure Statement

 (9) Declarations of Conformity and Summary Reports

(10) Device Description

(11) Executive Summary/Predicate Comparison

(12) Substantial Equivalence Discussion

(13) Proposed Labeling

(14) Sterilization and Shelf Life

(15) Biocompatibility

(16) Software

(17) Electromagnetic Compatibility and Electrical Safety

(18) Performance Testing – Bench

(19) Performance Testing – Animal

(20) Performance Testing – Clinical"

However, in addition to that, supplementary documents can be helpful. In particular, It is useful to attach the 510(k) Acceptance Checklist (RTA checklist[3]), which includes page numbers where each of the elements in the 510(k) can be found.

Note: You need to set reasonable expectations about 510(k) clearance. While FDA targets 90–100 calendar days for approval, you can expect one or two rounds of questions in a realistic case, which stops the regulatory timer. Thus, a six to nine months timeframe is much more real.

A.3 Substantial Equivalence and Predicate Devices

As we discussed earlier, Premarket Notification [510(k)] is a premarketing submission made to FDA to demonstrate that the device to be marketed is safe and effective by proving substantial equivalence (SE) to a legally marketed device. In order to demonstrate SE, the applicant must compare their device to a similar legally marketed US device(s). The legally marketed device to which equivalence is drawn is termed the predicate device.

---

[3]FDA, Acceptance Checklists for 510(k)s, Retrieved from https://www.fda.gov/medical-devices/premarket-notification-510k/acceptance-checklists-510ks

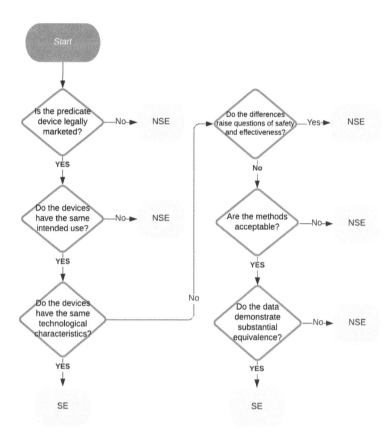

Figure A.2 510(k) decision-making flowchart.

A claim of substantial equivalence does not mean the device must be identical. Substantial equivalence is based on the intended use. It can be established with respect to design, energy used or delivered, materials, performance, safety, effectiveness, labeling, biocompatibility, standards, and other applicable characteristics.

In particular, the "substantial equivalence" claim can be made against a variety of devices. It can be any before 1976 or post-1976 device that **is** or **was** legally marketed in the United States. For example, an applicant may claim SE to a device that is no longer being marketed in the United States. However, SE cannot be claimed against a device found in violation of the Federal Food Drug & Cosmetic (FD&C) Act.

Thus, three important elements for establishing SE are (see Fig. A.2): (a) the predicate device is or was legally marketed in the United States, (b) the same intended use, and c) technological characteristics. If technical aspects are the same as a predicate device, it is a good base for establishing SE. If they are different (e.g., you are using a different physical principle), you need to prove that it is safe and efficient, methods are adequate, and performance data demonstrate substantial equivalence.

However, in most cases, a device recently cleared through 510(k) is used as a predicate device.

A.4 Summary vs. Statement

There are two forms of the 510(k) submission. The applicant can select between summary or statement submissions. The difference lies in the number of details, which will be available in the public domain.

Most applicants file summary submission. In this case, the summary document can be found through the 510(k) database. It contains information about the indications for use, the predicate device, the device description, comparison with the predicate device, and a summary of performance testing (e.g., standards tested against).

However, more information can be accessed by a third party by filing a request under the Freedom of Information Act (FOIA). In this case, the requestor may request the complete set of documents exchanged between the 510(k) submitter and the FDA. The FDA will redact from documents all proprietary information. However, a lot of information still will be visible.

To give the submitter more control over the availability of the disclosed information, the FDA provides an option to file a so-called statement submission. In this case, there is no information in the public domain. However, in exchange for that secrecy, the applicant promises to provide all information included in this Premarket Notification on safety and effectiveness to any requester in private communication within 30 days.

Appendix B

SaMD: Software as a Medical Device

Nowadays, almost every medical device relies on an ecosystem of software products. Some of them may fall under the definition of medical devices and can be a subject of regulatory scrutiny. Thus, early determination of whether it is a medical device can help guide efforts and deploy scarce resources accordingly. It could be helpful to understand the logic of regulators, especially for programmers who are not very familiar with these regulations. That is why we have included the criteria and examples of SaMD categorizations from the IMDRF Final document (IMDRF, "Software as a Medical Device:" Possible Framework for Risk Categorization and Corresponding Considerations, 2014[1]) in their entirely:

"7.3   Criteria for Determining SaMD Category
Criteria for Category IV:

   i. SaMD that provides information to treat or diagnose a disease or conditions in a critical situation or condition is a Category IV and is considered to be of very high impact.

Criteria for Category III:

   i. SaMD that provides information to treat or diagnose a disease or conditions in a serious situation or condition is a Category III and is considered to be of high impact.

   ii. SaMD that provides information to drive clinical management of a disease or conditions in a critical situation or condition is a Category III and is considered to be of high impact.

---

[1]IMDRF (2014). "Software as a Medical Device:" Possible Framework for Risk Categorization and Corresponding Considerations. Retrieved from http://www. imdrf.org/docs/imdrf/final/technical/imdrf-tech-140918-samd-framework-risk-categorization-141013.pdf

Criteria for Category II:

    i. SaMD that provides information to treat or diagnose a disease or conditions in a non-serious situation or condition is a Category II and is considered to be of medium impact.

    ii. SaMD that provides information to drive clinical management of a disease or conditions in a serious situation or condition is a Category II and is considered to be of medium impact.

    iii. SaMD that provides information to inform clinical management for a disease or conditions in a critical situation or condition is a Category II and is considered to be of medium impact.

Criteria for Category I:

    i. SaMD that provides information to drive clinical management of a disease or conditions in a non-serious situation or condition is a Category I and is considered to be of low impact.

    ii. SaMD that provides information to inform clinical management for a disease or conditions in a serious situation or condition is a Category I and is considered to be of low impact.

    iii. SaMD that provides information to inform clinical management for a disease or conditions in a non-serious situation or condition is a Category I and is considered to be of low impact.

7.4   Examples of SaMD

The examples below are intended to help illustrate the application of the framework and resulting categories.

Category IV:

SaMD that performs diagnostic image analysis for making treatment decisions in patients with acute stroke, i.e., where fast and accurate differentiation between ischemic and hemorrhagic stroke is crucial to choose early initialization of brain-saving intravenous thrombolytic therapy or interventional revascularization.

This example uses criteria IV.i from Section 7.3 in that the information provided by the above SaMD is used to treat a fragile patient in a critical condition that is life threatening, may require major therapeutic intervention, and is time sensitive.

SaMD that calculates the fractal dimension of a lesion and surrounding skin and builds a structural map that reveals the different growth patterns to provide diagnosis or identify if the lesion is malignant or benign.

This example uses criteria IV.i from Section 7.3 in that the information provided by the above SaMD is used to diagnose a disease that may be life threatening, may require major therapeutic intervention, and may be time sensitive.

SaMD that performs analysis of cerebrospinal fluid spectroscopy data to diagnose tuberculosis meningitis or viral meningitis in children.

This example uses criteria IV.i from Section 7.3 in that the information provided by the above SaMD is used to diagnose a disease in a fragile population with possible broader public health impact that may be life threatening, may require major therapeutic intervention, and may be time sensitive.

SaMD that combines data from immunoassays to screen for mutable pathogens/pandemic outbreak that can be highly communicable through direct contact or other means.

This example uses criteria IV.i from Section 7.3 in that the information provided by the above SaMD is used to screen for a disease or condition with public health impact that may be life threatening, may require therapeutic intervention and may be time critical.

Category III:
SaMD that uses the microphone of a smart device to detect interrupted breathing during sleep and sounds a tone to rouse the sleeper.

This example uses criteria III.i from Section 7.3 in that the information provided by the above SaMD is used to treat a condition where intervention is normally not expected to be time critical in order to avoid death, long-term disability or other serious deterioration of health.

SaMD that is intended to provide sound therapy to treat, mitigate or reduce effects of tinnitus for which minor therapeutic intervention is useful.

This example uses criteria III.i from Section 7.3 in that the information provided by the above SaMD is used to treat a condition that may be moderate in progression, may not require therapeutic intervention and whose treatment is normally not expected to be time critical.

SaMD that is intended as a radiation treatment planning system as an aid in treatment by using information from a patient and provides specific parameters that are tailored for a particular tumor and patient for treatment using a radiation medical device.

This example uses criteria III.ii from Section 7.3 in that the information provided by the above SaMD is used as an aid in treatment by providing enhanced support to the safe and effective use of a medical device to a patient in a critical condition that may be life threatening and requires major therapeutic intervention.

SaMD that uses data from individuals for predicting risk score in high-risk population for developing preventive intervention strategies for colorectal cancer.

This example uses criteria III.ii from Section 7.3 in that the information provided by the above SaMD is used to detect early signs of a disease to treat a condition that may be life-threatening disease impacting high-risk populations, may require therapeutic intervention and may be time critical.

SaMD that is used to provide information by taking pictures, monitoring the growth or other data to supplement other information that a healthcare provider uses to diagnose if a skin lesion is malignant or benign.

This example uses criteria III.ii from Section 7.3 in that the information provided by the above SaMD is used as an aid to diagnosing a condition that may be life-threatening, may require therapeutic intervention and may be time critical by aggregating relevant information to detect early signs of a disease.

Category II:
SaMD that analyzes heart rate data intended for a clinician as an aid in diagnosis of arrhythmia.

This example uses criteria from II.ii Section 7.3 in that the information provided by the above SaMD is used to aid in the diagnosis of a disease of a condition that may be moderate in progression, may not require therapeutic intervention and whose treatment is normally not expected to be time critical.

SaMD that interpolates data to provide 3D reconstruction of a patient's computer tomography scan image, to aid in the placement of catheters by visualization of the interior of the bronchial tree; in lung tissue; and placement of markers into soft lung tissue to guide radio surgery and thoracic surgery.

This example uses criteria II.ii from Section 7.3 in that the information provided by the above SaMD is used to aid in the next treatment intervention of a patient where the intervention is not normally expected to be time critical in order to avoid death, long-term disability, or other serious deterioration of health.

SaMD that uses data from individuals for predicting risk score for developing stroke or heart disease for creating prevention or interventional strategies.

This example uses criteria II.iii from Section 7.3 in that the information provided by the above SaMD is used to detect early signs of a disease to treat a condition that is not normally expected to be time critical in order to avoid death, long-term disability, or other serious deterioration of health.

SaMD that integrates and analyzes multiple tests utilizing standardized rules to provide recommendations for diagnosis in certain clinical indications, e.g., kidney function, cardiac risk, iron and anemia assessment.

This example uses criteria II.ii from Section 7.3 in that the information provided by the above SaMD is used to detect early signs of a disease to treat a condition that is no normally expected to be time critical in order to avoid death, long-term disability, orother serious deterioration of health.

Note: This example includes both serious and potentially non-serious conditions but per the categorization principle in Section 7.1 when a manufacturer's SaMD definition statement states that the SaMD can be used across multiple healthcare situations or condition it will be categorized at the highest category according to the SaMD definition statement.

SaMD that helps diabetic patients by calculating bolus insulin dose based on carbohydrate intake, pre-meal blood glucose, and anticipated physical activity reported to adjust carbohydrate ratio and basal insulin.

This example uses criteria II.ii from Section 7.3 in that the information provided by the above SaMD is used to aid in treatment of a condition not normally expected to be time critical in order to avoid death, long-term disability, or other serious deterioration of health.

Category I:
SaMD that sends ECG rate, walking speed, heart rate, elapsed distance, and location for an exercise-based cardiac rehabilitation patient to a server for monitoring by a qualified professional.

This example uses criteria I.ii from Section 7.3 in that the information provided by the above SaMD is an aggregation of data to provide clinical information that will not trigger an immediate or near term action for the treatment of a patient condition that is not normally expected to be time critical in order to avoid death, long-term disability, or other serious deterioration of health.

SaMD that collects data from peak-flow meter and symptom diaries to provide information to anticipate an occurrence of an asthma episode.

This example uses criteria I.ii from Section 7.3 in that the information provided by the above SaMD is an aggregation of data to provide best option to mitigate a condition that is not normally expected to be time critical in order to avoid death, long-term disability ,or other serious deterioration of health.

SaMD that analyzes images, movement of the eye or other information to guide next diagnostic action of astigmatism.

This example uses criteria I.i from Section 7.3 in that the information provided by the above SaMD is an aggregation of data to provide clinical information that will not trigger an immediate or near term action for the treatment of a patient condition that even if not curable can be managed effectively and whose interventions are normally noninvasive in nature.

SaMD that uses data from individuals for predicting risk score (functionality) in healthy populations for developing the risk (medical purpose) of migraine (non-serious condition.

This example uses criteria I.i from Section 7.3 in that the the information provided by the above SaMD is an aggregation of data to provide clinical information that will not trigger an immediate or near term action for the treatment of a patient condition that even if not curable can be managed effectively and whose interventions are normally noninvasive in nature.

SaMD that collects output from a ventilator about a patient's carbon dioxide level and transmits the information to a central patient data repository for further consideration.

This example uses criteria I. ii from Section 7.3 in that the information provided by the above SaMD is an aggregation of data to provide clinical information that will not trigger an immediate or near term action for the treatment of a patient condition that is not normally expected to be time critical in order to avoid death, long-term disability, or other serious deterioration of health.

SaMD that stores historical blood pressure information for a healthcare provider's later review.

This example uses criteria I.ii from Section 7.3 in that the information provided by the above SaMD is an aggregation of data to provide clinical information that will not trigger an immediate or near term action for the treatment of a patient condition that is not normally expected to be time critical in order to avoid death, long-term disability, or other serious deterioration of health.

SaMD intended for image analysis of body fluid preparations or digital slides to perform cell counts and morphology reviews.

This example uses criteria I.ii from Section 7.3 in that the information provided by the above SaMD is an aggregation of data to provide clinical information that will not trigger an immediate or near term action for the treatment of a patient condition that is not normally expected to be time critical in order to avoid death, long-term disability, or other serious deterioration of health.

SaMD intended for use by elderly patients with multiple chronic conditions that receives data from wearable health sensors, transmits data to the monitoring server, and identifies higher-level information such as tachycardia and signs of respiratory infections based on established medical knowledge and communicates this information to caregivers.

This example uses criteria I.ii from Section 7.3 in that the information provided by the above SaMD is an aggregation of data to provide clinical information that will not trigger an immediate or near term action for the treatment of a patient condition that is not normally expected to be time critical in order to avoid death, long-term disability, or other serious deterioration of health.

SaMD that uses hearing sensitivity, speech in noise, and answers to a questionnaire about common listening situations to self-assess for hearing loss.

This example uses criteria from I.ii Section 7.3 in that the information provided by the above SaMD is an aggregation of data to provide clinical information that will not trigger an immediate or near term action for the treatment of a patient condition that is not normally expected to be time critical in order to avoid death, long-term disability, or other serious deterioration of health."

Appendix C

Technology Readiness Levels in MedDev

The US Army Medical Research and Materiel Command (MRMC) developed TRL descriptions for the development of medical devices based on consideration of the applicable Food and Drug Administration (FDA) regulatory process along with industry practices and experience with their research and development (R&D) processes (discovery through manufacturing, production, and marketing). Their TRL is linked to "FDA events:"[1]

"TRL 1 – Lowest level of technology readiness. Maintenance of scientific awareness and generation of scientific and bioengineering knowledge base. Scientific findings are reviewed and assessed as a foundation for characterizing new technologies.

TRL 1 DECISION CRITERION: Scientific literature reviews and initial Market Surveys are initiated and assessed. Potential scientific application to defined problems is articulated.

TRL 2 – Intense intellectual focus on the problem with generation of scientific "paper studies" that review and generate research ideas, hypothesis, and experimental designs for addressing the related scientific issues.

TRL 2 DECISION CRITERION: Hypothesis(es) is generated. Research plans and/or protocols are developed, peer reviewed, and approved.

TRL 3 – Basic research, data collection, and analysis begin in order to test hypothesis, explore alternative concepts, and identify and evaluate component technologies. Initial tests of design concept,

---

[1]Wheatcraft, L. (2015). Technology Readiness Levels Applied to Medical Device Development. Retrieved from https://reqexperts.com/2015/11/30/technology-readiness-levels-applied-to-medical-device-development/

and evaluation of candidate(s). Study endpoints defined. Animal models (if any) are proposed. Design verification, critical component specifications, and tests (if a system component, or necessary for device Test and Evaluation) developed.

TRL 3 DECISION CRITERION: Initial proof-of-concept for device candidates is demonstrated in a limited number of laboratory models (may include animal studies).

TRL 4 – Non-good laboratory practice (GLP) laboratory research to refine hypothesis and identify relevant parametric data required for technological assessment in a rigorous (worst case) experimental design. Exploratory study of candidate device(s)/systems (e.g., initial specification of device, system, and subsystems). Candidate devices/systems are evaluated in laboratory and/or animal models to identify and assess potential safety problems, adverse events, and side effects. Procedures and methods to be used during non-clinical and clinical studies in evaluating candidate devices/systems are identified. The design history file, design review, and when required a master device record, are initiated to support either a Premarket Notification (510(k)) or PMA for medical devices.

TRL 4 DECISION CRITERION: Proof-of-concept and safety of candidate devices/systems demonstrated in defined laboratory/ animal models.

TRL 5 – Further development of selected candidate(s). Devices compared to existing modalities and indications for use and equivalency demonstrated in model systems. Examples include devices tested through simulation, in tissue or organ models, or animal models if required. All component suppliers/vendors are identified and qualified; vendors for critical components audited for Current Good Manufacturing Practices (cGMP)/ Quality System Regulation (QSR) compliance. Component tests, component drawings, design history file, design review, and any master device record verified. Product Development Plan drafted. Pre-Investigational Device Exemption (IDE) meeting held with Center for Devices and Radiologic Health (CDRH) for proposed Class III devices, and the IDE is prepared and submitted to CDRH. For a 510(k), determine substantially equivalent devices and their classification, validate functioning model, ensure initial testing is complete, and validate data and readiness for cGMP inspection.

TRL 5 DECISION CRITERION: IDE review by CDRH results in determination that the investigation may begin. For a 510(k), preliminary findings suggest the device will be substantially equivalent to a predicate device.

TRL 6 – Clinical trials conducted to demonstrate safety of candidate Class III medical device in a small number of humans under carefully controlled and intensely monitored clinical conditions. Component tests, component drawings, design history file, design review, and any master device record updated and verified. Production technology demonstrated through production-scale cGMP plant qualification. For 510(k), component tests, component drawings, design history file, design review, and any master device record updated and verified. Manufacturing facility ready for cGMP inspection.

TRL 6 DECISION CRITERION: Data from the initial clinical investigation demonstrate that the Class III device meets safety requirements and supports proceeding to clinical safety and effectiveness trials. For a 510(k), information and data demonstrate substantial equivalency to predicate device and support production of the final prototype and final testing in an operational environment.

TRL 7 – Clinical safety and effectiveness trials conducted with a fully integrated Class III medical device prototype in an operational environment. Continuation of closely controlled studies of effectiveness, and determination of short-term adverse events and risks associated with the candidate product. Functional testing of candidate devices completed and confirmed, resulting in final down-selection of prototype device. Clinical safety and effectiveness trials completed. Final product design validated, and final prototype and/or initial commercial-scale device are produced. Data collected, presented, and discussed with CDRH in support of continued device development. For a 510(k), final prototype and/or initial commercial-scale device are produced and tested in an operational environment.

TRL 7 DECISION CRITERION: Clinical endpoints and test plans agreed to by CDRH. For a 510(k), information and data demonstrate substantial equivalency to predicate device and use in an operational environment, and support preparation of 510(k).

TRL 8 – Implementation of clinical trials to gather information relative to the safety and effectiveness of the device. Trials are conducted to evaluate the overall risk-benefit of using the device and to provide an adequate basis for product labeling. Confirmation of QSR compliance, the design history file, design review, and any master device record, are completed and validated, and device production followed through lot consistency and/or reproducibility studies. Pre-PMA meeting held with CDRH. PMA prepared and submitted to CDRH. Facility PAI (cGMP/QSR/Quality System Regulation (QSIT)) completed. For 510(k), prepare and submit application.

TRL 8 DECISION CRITERION: Approval of the PMA [or, as applicable, 510(k)] for device by the CDRH.

TRL 9 – The medical device may be distributed/marketed. Post marketing studies (non-clinical or clinical) may be required and are designed after agreement with the FDA. Post marketing surveillance.

TRL 9 DECISION CRITERION: None – continue surveillance"

Appendix D

Useful Resources

As we mentioned through the book, MedTech entrepreneurship is a long and enduring process. It is better to do in a good company. In addition to your co-founders, there are several companies, which spread knowledge and make life easier.

Below, one can see a curated list of companies and resources, which I found particularly helpful for MedTech entrepreneurs (Full disclosure: I am not affiliated with any of these companies).

Product Development

StarFish Medical (starfishmedical.com) is a MedDev engineering boutique headquartered in Victoria, BC. It also has a second location in Toronto, ON. StarFish Medical runs a Medical Device Playbook conference, which brings many industry professionals in one place. The company also has numerous helpful resources on its website.

Regulatory Affairs

EMERGO (www.emergobyul.com) is a global MedDev consulting company. It is a subsidiary of UL. In addition to their paid services, they have multiple free resources, including lists of MedDev regulatory documents per country.

Reimbursement Affairs

MCRA (mcra.com) is a leading Medical Device and Biologics CRO with fully integrated regulatory, reimbursement, and compliance services. They regularly provide webinars on various topics, including reimbursement affairs.

Funding and Partnering

Life Science Nation (www.lifesciencenation.com) is a Boston-based company highly visible in the life science space. Their flagship product is a series of RESI conferences (resiconferences.com), run in several cities across the United States. The conferences contain partnering events that can be particularly useful for MedTech entrepreneurs. Note that each city has a slightly different flavor of the audience. For example, Boston is tilted more toward pharma. So, some preliminary search and optimization may be required to achieve the maximum effect.

Life Science Nation also runs multiple webinars on a variety of topics, including non-dilutive funding.

Index

Milton Keynes UK
Ingram Content Group UK Ltd.
UKHW051533141024
449569UK00001B/12